Advances in Aquatic Ecology

— Volume 4 —

Advances in Aquatic Ecology

— Volume 4 —

Editor
Dr. Vishwas B. Sakhare
Head,
Post Graduate Department of Zoology
Yogeshwari Mahavidyalaya,
Ambajogai – 431 517
Maharashtra

2010
DAYA PUBLISHING HOUSE
Delhi - 110 035

ISBN 81-7035-657-1
ISBN 978-81-7035-657-8

Published by : **Daya Publishing House**
1123/74, Deva Ram Park
Tri Nagar, Delhi - 110 035
Phone: 27383999
Fax: (011) 23260116
E-mail: dayabooks@vsnl.com
website: www.dayabooks.com

Showroom : 4760-61/23, Ansari Road, Darya Ganj,
New Delhi - 110 002
Phone: 23245578, 23244987

Laser Typesetting : **Classic Computer Services**
Delhi - 110 035

Printed at : **Chawla Offset Printers**
Delhi - 110 052

PRINTED IN INDIA

Dedicated

to

My Reverend Guide

Dr. P.K. Joshi

Reader in Zoology

Dnyanopasak Mahavidyalaya, Parbhani

Maharashtra

Preface

I am delighted to write about the fourth volume of *Advances in Aquatic Ecology*. This volume is the compilation of esteemed articles of internationally acknowledged experts in the field of aquatic ecology with the intention of providing a sufficient depth of the subject to satisfy the need of a level which will be comprehensive and intresting.It is an assemblage of up to date information of rapid advances and developments taking place in the field of aquatic ecology. With its application oriented and interdisciplinary approach, I hope that the students, teachers, researchers, scientists, policy makers and environmental lawyers in India and abroad will find this volume much more useful. The articles in the book have been contributed by eminent scientists/academicians active in the field of aquatic ecology.

My special thanks and appreciation go to the scientists whose contributions have enriched this volume. I wish to express my sincere gratitude to Dr. Sureshji Khursale, President, Yogeshwari Education Society, Ambajogai who has been a source of constant inspiration .Iam especially thankful to Dr.Prakash Prayag, Principal, Yogeshwari Mahavidyalaya, Ambajogai for his encouragement. I owe my special thanks to Dr. P.K. Joshi and Dr. S.P. Chavan of Dnyanopasak Mahavidyalaya, Parbhani, Dr. Mohan S. Kodarkar of Indian Association of Aquatic Biologists, Hyderabad, Dr. Indranil Ghosh of West Bengal University of Animal and Fishery Sciences, Kolkata, Dr. Milind Girkar of College of Fisheries, Udgir, Dr. Meenakshi Jindal of CCS Haryana Agricultural University, Hisar, Dr. Vishwas Shembekar of Rajarshi Shau College, Latur, Shri Sachin Satam of MPEDA, Mumbai, Prof. M.B. Mule of Dr. Babasaheb Ambedkar Marathwada University,Aurangabad, Prof. Ravishankar Piska of Osmania University, Hyderabad, Prof. S. Ajmal Khan, Dr. K. Sivakumar and G. Thirumaran of Annamalai University, Parangipettai and Prof. A.K. Patra of Utkal University, Bhubaneshwar.

I wish to thank my wife Surekha for her endurance during the compilation of work of this volume. She has helped me for constantly data feeding, processing and reprocessing on computer. I want to thank my father Shri Balasaheb Sakhare, brother Avinash and sisters Minal and Preeti for their help in many ways. I thank my publisher Shri Anil Mittal of Daya Publishing House, Delhi for taking pains in bringing out the book.

Finally, I will always remain a debtor to all my well-wishers for their blessings, without which this volume would not have come into existence.

Dr. Vishwas Balasaheb Sakhare

Contents

List of Contributors

Anantharaman, P.
CAS in Marine Biology, Annamalai University, Parangipettai - 608 502, Tamil Nadu

Arumugam, R.
CAS in Marine Biology, Annamalai University, Parangipettai – 608 502, Tamil Nadu

Asimi, O.A.
Faculty of Fisheries, S.K. University of Agricultural Sciences and Technology of Kashmir, Kashmir – 190 006, Jammu & Kashmir

Babare, M.G.
P.G. Department of Zoology/Fishery Science, A.S.C.College, Naldurg – 413 602, Maharashtra

Chaudhari, B.S.
District Project Facilator, District Agriculture Office, Osmanabad District, Maharashtra

Chavan, S.P.
P.G. Department of Zoology, Dnyanopasak Mahavidyalaya, Parbhani – 431 401, Maharashtra

Dande, K.G.
Department of Zoology and Dairy Science, Mahatma Basweshwar College, Latur – 413 512, Maharashtra

Devi, Achyutha J.
Department of Zoology, University College of Science, Osmania University, Hyderabad – 500 007, Andhra Pradesh

Devi Suneetha, D.
Department of Zoology, University College of Science, Osmania University, Hyderabad – 500 007, Andhra Pradesh

Dutta (Roy) Sarbani
Fisheries and Aquaculture Extension Laboratory, Department of Zoology, University of Kalyani, Kalyani – 741 235, West Bengal

Gawade, Mangesh
College of Fisheries, Shirgaon, Ratnagiri, Maharashtra

Geetha, K.
Department of Zoology, University College of Science, Osmania University,Hyderabad – 500 007, Andhra Pradesh

Ghorpade, B.N.
Shankarrao Mohite Mahavidyalaya, Akluj – 413 101, Maharashtra

Ghosh, Indranil
Department of Aquaculture, West Bengal University of Animal and Fishery Science, Kolkata, West Bengal

Girkar, M.M.
Assistant Professor, Department of Fishery Hydrography, College of Fishery Science, Udgir – 413 517 Maharashtra

Harney, N.V.
Nilkanthrao Shinde Science and Arts College, Bhadrawati – 442 902, Maharashtra

Jain, K.L.
Department of Zoology and Aquaculture, CCS HAU, Hisar – 125 004, Haryana

Jindal, Meenakshi
Department of Zoology and Aquaculture, CCS HAU, Hisar – 125 004, Haryana

Jyothirmai, G.S.
Department of Zoology, University College of Science, Osmania University, Hyderabad – 500 007, Andhra Pradesh

Khabade, S.A.
Department of Zoology, D.K.A.S.C. College, Ichalkaranji, Maharashtra

Kulkarni, D.A.
Shikshan Maharashi Dnyandeo Mohekar Mahavidyalaya, Kallam – 413 507, Maharashtra

Kumbhar, A.C.
Shankarrao Mohite Mahavidyalaya, Akluj – 413 101, Maharashtra

Lokhande, M.V.
Department of Zoology and Dairy Science, Mahatma Basweshwar College, Latur – 413 512, Maharashtra

Manjusha, P.
Department of Zoology, University College of Science, Osmania University, Hyderabad – 500 007, Andhra Pradesh

Maqsood, Sajid
Faculty of Fisheries, Sher-e-Kashmir University of Agricultural Science and Technology–Kashmir, Jammu & Kashmir

Mote, M.V.
P.G. Department of Zoology / Fishery Science, A.S.C.College, Naldurg – 413 602
Maharashtra

Mule, M.B.
Department of Environmental Science, Dr. Babasaheb Ambedkar Marathwada University, Aurangabad – 431 004, Maharashtra

Nasare, P.N.
Nilkanthrao Shinde Science and Arts College, Bhadrawati – 442 902, Maharashtra

Niture, S.D.
Department of Zoology, J.E.S. College, Jalna – 431 203, Maharashtra

Panigrahi, A.K.
Fisheries and Aquaculture Extension. Laboratory, Department of Zoology, University of Kalyani, Kalyani – 741 235, West Bengal

Piska, Ravi Shankar
Department of Zoology, University College of Science, Osmania university, Hyderabad – 500 007, Andhra Pradesh

Rajagopal, S.
Centre of Advanced Study in Marine Biology, Annamalai University, Parangipettai – 608 502, Tamil Nadu

Rathod, D. S .
Department of Zoology and Fishery Science, Rajarshi Shahu College, Latur – 413 512, Maharashtra

Roy, D.
Reader, Department of Zoology, S.M.M. Town Post Graduate College, Ballia – 277 001, U.P.

Salunke, S.
Karmaveer Bhaurao Patil Mahavidyalaya, Pandharpur, Maharashtra

Samoon, M.H.
Faculty of Fisheries, Sher-e-Kashmir University of Agricultural Science and Technology – Kashmir, Jammu & Kashmir

Saravanan
Centre of Advanced Study in Marine Biology, Annamalai University, Parangipettai – 608 502, Tamil Nadu

Satam, S.B.
Marine Products Export Development Authority (MPEDA), Mumbai, Maharashtra

Sawant, Gauri
College of Fisheries, Shirgaon, Ratnagiri – 415629, Maharashtra

Shirdhankar, M.M.
College of Fisheries, Shirgaon, Ratnagiri – 415 629, Maharashtra

Singh Prabjeet
Faculty of Fisheries, Sher-e-Kashmir University of Agricultural Science and Technology, Kashmir – 190 006, Jammu and Kashmir

Shah, T.H.
Faculty of Fisheries, S.K. University of Agricultural Sciences and Technology of Kashmir, Kashmir – 190 006, Jammu & Kashmir

Shembekar, V.S
Department of Zoology and Fishery Science, Rajarshi Shahu College, Latur – 413 512, Maharashtra

Simmi
Department of Zoology and Aquaculture, CCS HAU, Hisar – 125 004, Haryana

Sitre, S.R.
Nilkanthrao Shinde Science and Arts College, Bhadrawati – 442 902, Maharashtra

Srivastava, P.K.
Department of Zoology, S.M.M. Town Post Graduate College, Ballia – 277 001, U.P.

Srivastava, P.P.
National Bureau of Fish Genetic Resources, Telibagh, Lucknow, Uttar Pradesh

Thakare, Abhijeet
College of Fisheries, Shirgaon, Ratnagiri – 415 629, Maharashtra

Tandale, A.T.
Department of Fish Processing Technology & Microbiology, College of Fishery Science, Udgir – 413 517, Maharashtra

Thirumaran, G.
CAS in Marine Biology, Annamalai University, Parangipettai – 608 502, Tamil Nadu

Todkari, S.S.
In charge, Quality Control, Karunya Marine Exports Pvt. Ltd., Ratnagiri – 415 612, Maharashtra

Verma, R.K.
Department of Zoology and Aquaculture, CCS HAU, Hisar – 125 004, Haryana

Vijayanand, P.
Centre of Advanced Study in Marine Biology, Annamalai University, Parangipettai – 608 502, Tamil Nadu

Vyas, Vipin
Department of Limnology, Barkatullah University, Bhopal – 462 026, Madhya Pradesh

Wadhave, N.S.
Nilkanthrao Shinde Science and Arts College, Bhadrawati – 442 902, Maharashtra

Chapter 1

Effect of Fluoride on Tissue Glycogen Levels in Freshwater Catfish, *Clarias batrachus* (Linn)

☆ *Achyutha Devi. J. and Ravi Shankar Piska*

Introduction

Clarias batrachus is fresh water euryhaline and eurythermal air breathing fish. It is the cheapest source of protein for millions of Indians. Fluoride enters into aquatic environment including municipal wastewaters and effluents from fertilizer producing plants and aluminium refineries (Giguere and Campbell 2004) Though, great advances have been made in fluoride research there remains much to be done on fresh water models on which the reports are sporadic, therefore, the present study was undertaken to investigate the effect fluoride on glycogen levels of the fresh water catfish, *Clarias batrachus.*

Materials and Methods

Fishes weighing about 75g to 100g were collected from Kaikaluru, Krishna District and acclimatized as suggested by Klontz and Smith (1969) for approximately 20 days, the experimental tubs were setup with paraliel controls.

Fish were exposed to 1,10,20, and 30ppm of Sodium Fluoride to different exposure spans of 1,30, and 60 days. Fish were dissected and the liver, kidney, muscle, gill, and brain used for estimation of glycogen level by the method of Nacholas *et al.* (1956). Statistical tests were used as per Bailey (1959).

Results

The tissue glycogen content in all the experimental tissues was found decreased throughout the exposure span. Time and concentration dependent reduction of glycogen was observed in all the tissues. Maximum amount of depletion was observed on the 60th day of exposure. Depletion was found maximum in kidney (–32.87 per cent P<0.05) followed by brain (–27.91 per cent P<0.001), liver (–27.86 per cent P<0.001), muscle (–24.79 per cent P<0.001) and gill (–10.65 per cent P<0.001). However slight fluctuations were observed on 60th day of exposure at lower concentrations of fluoride. Glycogen values of all tissues showed statistical significance over control throughout the exposure span except during 1st day of exposure, where in the values of muscle, kidney, gill, showed statically insignificant values at 1 ppm.

Table 1.1

Fluoride Concentration	*1 Day*		*30 Days*		*60 Days*	
	Mean ± SE	*Per cent V*	*Mean ± SE*	*Per cent V*	*Mean ± SE*	*Per cent V*
			Liver			
Control	2.94±0.02		3.01± 0.01		3.20±0.03	
1ppm	2.82 ± 0.01**	–4.08	2.75 ± 0.01*	–8.66	2.88±0.02*	–10
10ppm	2.66±0.03*	–9.52	2.65±0.02*	–14.98	2.66±0.015*	–16.87
20ppm	2.74±0.03*	–6.8	2.38±0.03*	–20.56	2.51±0.04**	–21.56
30ppm	2.54±0.01*	–13.61	2.29±0.04*	–23.45	2.31±0.03*	–27.86
			Muscle			
Control	4.51±0.05		4.6±0.08		4.7±0.09	
1 ppm	4.38 ± 0.06NS	–2.88	4.09±0.09**	–11.09	4.12±0.05*	–12.71
10 ppm	4.26±0.07***	–5.54	3.88±0.06*	–15.65	3.77±0.09*	–20.12
20 ppm	4.11±0.06*	–83.87	3.69±0.05*	–19.75	3.46±0.08*	–20.69
30 ppm	4.04±0.06*	–10.42	3.54±0.08*	–23.04	3.35±0.08*	–24.79
			Kidney			
Control	2.76±0.02		2.82±0.03		3.01±0.03	
1 ppm	2.67±0.03NS	–3.26	2.59±0.02*	–8.16	2.67±0.03**	–12.29
10 ppm	2.58±0.01*	–6.52	2.41±0.03*	–14.54	2.4±0.04*	–2.27
20 ppm	2.45±0.02*	–11.23	2.27±0.02*	–19.5	2.26±0.05*	–24.92
30 ppm	2.5±0.04*	–9.42	2.18±0.03*	–22.69	2.02±0.05*	–32.89
			Gill			
Control	0.98±0.01		0.98±0.01		0.98±0.01	
1 ppm	0.97±0.01NS	–0.4	0.99±0.02*	–0.43	0.97±0.02*	–1.02
10 ppm	0.95±0.02*	–3.71	0.94±0.02*	–3.69	0.87±0.02*	–10.65
20 ppm	0.96±0.02*	–2.15	0.91±0.01*	–6.45	0.88±0.02*	–9.83
30 ppm	0.93±0.02*	–4.4	0.9±0.02*	–7.78	0.85±0.02*	–9.8

Contd...

Table 1.1–Contd...

Fluoride Concentration	1 Day		30 Days		60 Days	
	Mean ± SE	Per cent V	Mean ± SE	Per cent V	Mean ± SE	Per cent V
			Brain			
Control	2.34±0.05		2.46±0.02		2.58±0.02	
1 ppm	2.29±0.04*	–2.14	2.11±0.02*	–14.23	2.32±0.01*	–10.08
10 ppm	2.18±0.03***	–6.84	2.25±0.03*	–8.34	2.14±0.03*	–17.05
20 ppm	2.11±0.02*	–9.83	2.06±0.06*	–10.26	1.99±0.02*	–22.87
30 ppm	2.09±0.01*	–10.6	2.01±0.05*	–18.29	1.86±0.03*	–27.91

(Values are mean ± SE of six observations per cent V= per cent variation from control values are significant at *p<0.001** p<0.01***p<0.05, NS = not significant).

Discussion

Glycogen a reserve energy source, is found to decrease during the exposure to fluoride. In animals glycogen is generally stored in liver, and muscle tissues sometimes also in kidney and brain. In the present investigations, there is a general trend of depletion in glycogen levels in all the tissues throughout the exposure span. The maximum decrease in glycogen level was found in kidney. It is presumable that fluoride disrupts carbohydrate metabolism and kidney assumes greater importance because of its nephrotoxicity. The nephrotoxic nature of fluoride was reported by (Spencer, 1980).

Interference with carbohydrate metabolism is frequently noted following acute fluoride intoxication (Kumar *et al.*, 2007) Fluoride poisoning induced dramatic changes in carbohydrate metabolism, which manifest themselves as a rapid decline in glycogen levels in liver, muscle, kidney, brain, and gill. The marked changes in the tissue glycogen concentration produced by varying doses of fluoride 1,10,20,and 30ppm indicate enzyme inhibition of fluoride.

Fluorides are recognized as strong inhibitors of enzymes associate with glycolysis. The enzyme enolase present in all tissues is a dehydrogenase involved in the anaerobic metabolism of carbohydrates. Loss of glycogen in the fluoride treated animals indicated that it caused glycogen breakdown. Enzyme enolase is activated by magnesium and in the presence of a phosphate buffer, it is inhibited by fluoride with the formation of magnesium fluorophosphates (Cannell, 1960). Loss of glycogen indicated the inhibition synthesis under stress conditions. Handler *et al.* (1946) claim that fluoride poisoning is also accompanied by depletion of liver and muscle glycogen and elevated blood lactate. The decline in the glycogen suggests enhanced conversion of glycogen to glucose to meet an increased energy requirements. Kasturi and Chandran (1997) have also made a similar suggestion in their study with *Mystus gulio* exposed to lead..

Similar observations were also reported in fry of *Catla catla* exposed to fluorine emitting industry effluents (Pillai and Mane 1984). Fluoride intoxication on the liver and muscle of mice (Singh, 1984) and in rats (McGown and Suttie, 1977). These observations extend substantial support to the present study. Decreasing glycogen content has been attributed to accumulation of citrate and lowering of adenosine triphosphate in tissue due to blockade in tricarboxylic cycle at citrate state. Hyperglycemia may result from many factors in addition to hepatic glycogenolysis including gluconeogenesis and lactate mobilization from muscle and inhibition enzyme enolase. Decreased glycogenesis or neoglycogenesis may be attributed to decreased glycogen content in liver. Rapid decrease in the glycogen levels was reported in fishes during stress and strarvation (Swallow and Flemming, 1969).

The general decrease in the glycogen levels was observed in different species of fish might be due to the onset of hypoxic conditions, due to decreased rate of oxygen consumption was reported in fresh water field crab after heavy metal intoxication (Reddy, 1981).

HMP shunt is an alternate pathway to glycolysis and TCA cycle for the oxidation of glucose. Fluoride stimulates oxidative metabolism has been demonstrated by an enhanced monophosphate shunt activity (Sbarra *et al.*, 1959) supports the present study.

Conclusion

The depletion of glycogen in kidney, brain, liver, muscle and gill of fluoride treated fish is due to rapid hepatic glycogenolysis, lactate mobilization from muscle and inhibition of enzyme enolase.

References

Bailey, N.T.J., 1954. *Statistics in Biology*. The ELBS English University Pub.

Cannel, WA., 1960. *Medical and Dental Aspects of fluoridation*. KK Lewis and Co. Ltd., London, p. 47–60.

Chitra, T. and Ramana Rao, J.V., 1982. Biochemical variations in *Channa punctatus* (Bloch) due to sodium Fluoride treatment. *Fluoride*, 13(2): 15–23.

Gigures, A. and Campbell, P.C.G., 2004. Fluroide toxicity towards fresh water organisms and hardness effects review and reanalysis of existing data. *Rev. Sci. Eau.*, 17(3): 373–393.

Handler, P., Herring, H.E. Jr. and Hebb, H., 1946.The effect of insulin in fluoride poison rats. *J. Biol. Chem.*, 164: 679–683

Kasturi, J and Chandfan, M.R., 1997. Sublethel effect of lead on feeding energetic, growth performance, bio chemical composition and accumulation of the estuarine catfish, *Mystus gulio* (Hamilton). *J. Environ. Biol.*, 18(1): 95–101.

Kumar, A., Nalini Tripati and Madhu Tripati, 2007. Flouoride induced biochemical changes in fresh water catfish *Clarias batrachus* (Linn). *Fluroide*, 40(1): 37–41.

McGown, E.L. and Suttie, J.W., 1977. Mechanism of Fluoride induced hyperglycemia in the rat. *Toxicol. Appl. Pharmacol.*, 40: 83–90.

Nacholas, V., Robert, C., Longley, W. and Joseph, H.R., 1956. Determination of glycogen in liver and muscle by using anthrone reagent. *J. Biochem.*, 22: 583–587.

Pillai, K.S. and Mane, U.H., 1984. Effect of Fluoride effluents on some metabolites and minerals in fry of *Catla catla* (Hamilton). *Fluoride*, 17(4) Oct.

Reddy, S.L.N., Shankariah, K. and Ramana Rao, J.V., 1988. Changes in carbohydrate metabolism in a fresh water field crab during Fluoride toxicity. *American Society of Pharmacology and Experimental Therapeutics/Society of Toxicology Conference.*

Reddy, S.L.N., 1981. Physiological responses of *Barytelpusa guerini* to environmental variation (pollution). *Ph.D. Thesis*, Osmania University, Hyderabad.

Sbarra, A.J. and Karovsky, M.L., 1959. The biochemical basis of phagocytosis. I. Metabolic changes during the ingestion of particles by polymorphonuclear leucocytes. *J. Biol. Chem.*, 234: 1355.

Singh, M., 1984. Biochemical and cytochemical alterations in liver and kidney following experimental Fluorosis. *Fluoride*, 17(2):81–83.

Spencer, H., Kramer, L., Gatza, Norris, Wiatrowski, C. and Gandhi, V.C., 1980 Fluoride metabolism in patients with chronic renal failure. *Arch. Int. Med.*, 140: 1331–1335.

Swollow, Richard and Flemming, W.R., 1969. The effect of starvation feeding,glucose and ACTH on the liver glycogen levels of *Tilapia mossambica*. *Comp. Biochem. Physiol.*, 28(1): 95–106.

Chapter 2

Effect of Three Different Doses of Dried Water Hyacinth and Poultry Manure in the Production of Cladocerans

☆ *P.K. Srivastava and D. Roy*

Introduction

Success in culturing planktivorous fish depends primarily on zooplankton density (Fernando, 1994). Organic manures contain all essential ingredients to promote plankton density (Arun Kumar, 2005). Inadequate zooplankton in nursery pond results in high larval mortality (Alikunhi *et al.*, 1955). Doses of organic manures in culture system and physico-chemical properties of water play a significant role on density of cladocerans (Delbaere and Dhert, 1996).

Daphnia carinata, *Moina micrura* and *Ceriodaphnia cornuta* belong to order Cladocera, which are small crustaceans popularly known as 'water fleas'. Cladocerans are easily digestible due to the presence of digestive enzymes (Kumar *et al.*, 2005) and their abundance holds the clue to successful rearing of Indian major carp and exotic carp during early ontogeny (Sharma and Chakrabarti, 2000 and Kumar *et al.*, 2000). Cladocerans have been cultured by using rice bran (De Pauw *et al.*, 1981), soybean and wheat bran (Lee, 1982), poultry manure (Delbaere and Dhert, 1996) dry cow dung and garden soil (Singh and Dutta Munshi, 1991), Pig and chicken manure (Tay, 1980) and cow dung, mustard oil cake and poultry manure (Jana and Chakrabarti, 1993). Mitra and Banerjee (1976); Keke *et al.* (1994) and Sharma (2003) have used water hyacinth as organic manure and found better zooplankton production. Eradication of water hyacinth from water bodies is a problem throughout India and being a menace, water hyacinth was selected to work out its impact on production of live food so as to contribute in its eradication. Little attention has been paid to evaluate impact of water hyacinth on the production of live food and further study with water hyacinth in combination with other organic manure is needed to make this weed more effective for the production of zooplankton.

The objective of the present study is to investigate effect of three different doses of water hyacinth and poultry manure on the production of *Daphnia carinata, Moina micrura* and *Ceriodaphnia cornuta* and to evaluate whether these doses exert any adverse effect on physico-chemical properties of water. The present investigation is an attempt to find out optimum manure dose, which will be used to obtain higher production of live food for larval rearing as well as in polyculture system.

Material and Methods

Zooplankton organisms (*D. carinata, Moina micrura* and *Ceriodaphnia cornuta*) were collected by plankton net of 53 mm mesh size from Surha lake (25°48′ and 25°52′ N Lat and 84°8′ and 84°13′ E Long) located in Ballia District in Uttar Pradesh (India).

Preparation of Culture Tanks

Mass culture of *D. carinata, M. micrura* and *C. cornuta* were carried out separately in circular cemented tanks (each with 100 litres water holding capacity) maintained in outdoor condition where the environmental conditions were same for all the tanks. Cemented tanks were cleaned and filled with tap water up to 100 Liter and were left for four days for dechlorination.

Manuring and Inoculation

A mixture of organic manure- dried water hyacinth (DWH) and poultry manure (PM) were applied in three different doses in the ratio of 1:1- first dose- 1.26 $g.l^{-1}$, second dose- 2.52 $g.l^{-1}$ and third dose- 5.04 $g.l^{-1}$. Water hyacinth was harvested, sun dried and crushed into small pieces before its application. Organic manures were allowed to decompose for 10 days. Inoculation @10 $inds.l^{-1}$ was done on 11th day after fertilization. Three replicates were used for each treatment.

Quantitative Analysis of Zooplankton

Samples of zooplankton were collected every alternate day from 4th day after inoculation up to the 30th day. Collected samples were fixed with 4 per cent formalin for subsequent microscopic quantitative analysis. One ml of fixed sample was placed in Sedgwick-Rafter plankton counting cell (50mm O 10mm O 1mm). The number of zooplankton organisms was expressed as individuals per litre ($inds.l^{-1}$).

Physico-chemical Parameters

Water quality of cultured tank water was monitored regularly by collecting water sample at 10 am. Temperature, pH, dissolved oxygen, free carbon dioxide, ammonia, nitrate and phosphate were analyzed according to APHA (1995).

Statistical Analysis

Zooplankton population and physico-chemical parameters were statistically compared by one-way analysis of variance (one-way ANOVA).

Results

Quantitative Analysis of Zooplankton

During entire period of investigation, in first (1.26 $g.l^{-1}$), second (2.52 $g.l^{-1}$) and third (5.04 $g.l^{-1}$) doses population of *D. carinata* ranged from 19-243, 52-639 and 37-765 $inds.l^{-1}$ respectively

(Figure 2.1.), population of *M. micrura* varied from 26-392, 63-527 and 54-923 inds.l^{-1} respectively (Figure 2.2.) and population of *C. cornuta* ranged from 72-974, 165-1893 and 118-2151 inds.l^{-1} respectively (Figure 2.3.).

In first dose peak population of *D. carinata* (243 inds.l^{-1}) was found on 8th day of inoculation whereas peak population of *M. micrura* (392 inds.l^{-1}) and *C. cornuta* (974 inds.l^{-1}) was observed on 12th day. In second dose peak population of *D. carinata* (639 inds.l^{-1}) and *M. micrura* (527 inds.l^{-1}) was found on 16th day and *C. cornuta* (1893 inds.l^{-1}) on 18th day. In third dose peak population of *M. micrura* (923 inds.l^{-1}) appeared on 18th day whereas peak population of *D. carinata* (765 inds.l^{-1}) and *C. cornuta* (2151 inds.l^{-1}) on 20th day.

In terms of average population, number of *D. carinata* was 3.726 times higher in third dose and 2.759 times higher in second dose than first dose. Similarly, number of *M. micrura* was 2.714 times higher in third dose and 1.564 times higher in second dose than first dose and number of *C. cornuta* was 3.174 times higher in third dose and 2.608 times higher in second dose than first dose. The number of zooplankton organisms decreased faster in first dose than the other two higher doses.

Statistically, the population density of *D. carinata, M. micrura* and *C. cornuta* was significantly ($P<0.01$) affected by the different doses of manure as well as the culture durations. Among three doses, significantly ($P<0.01$) higher number of individuals was observed in the third dose followed by second and first dose.

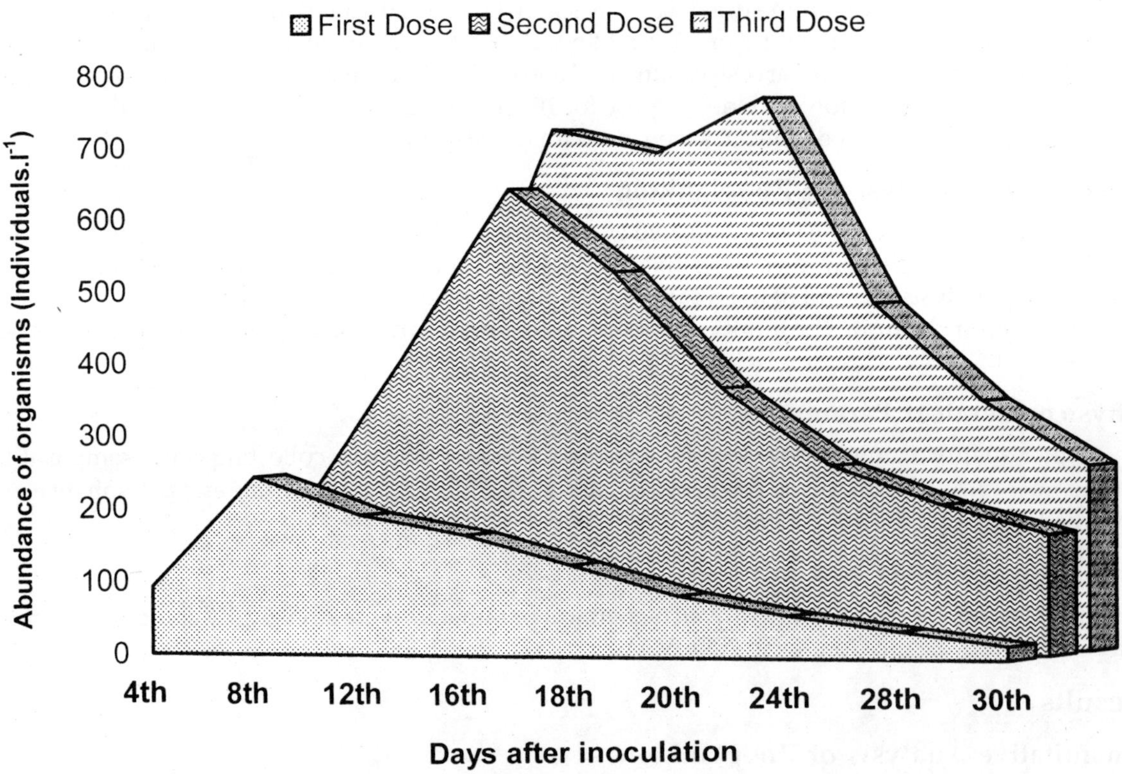

Figure 2.1: Abundance of *Daphnia carinata* in Three Doses of DWH and PM

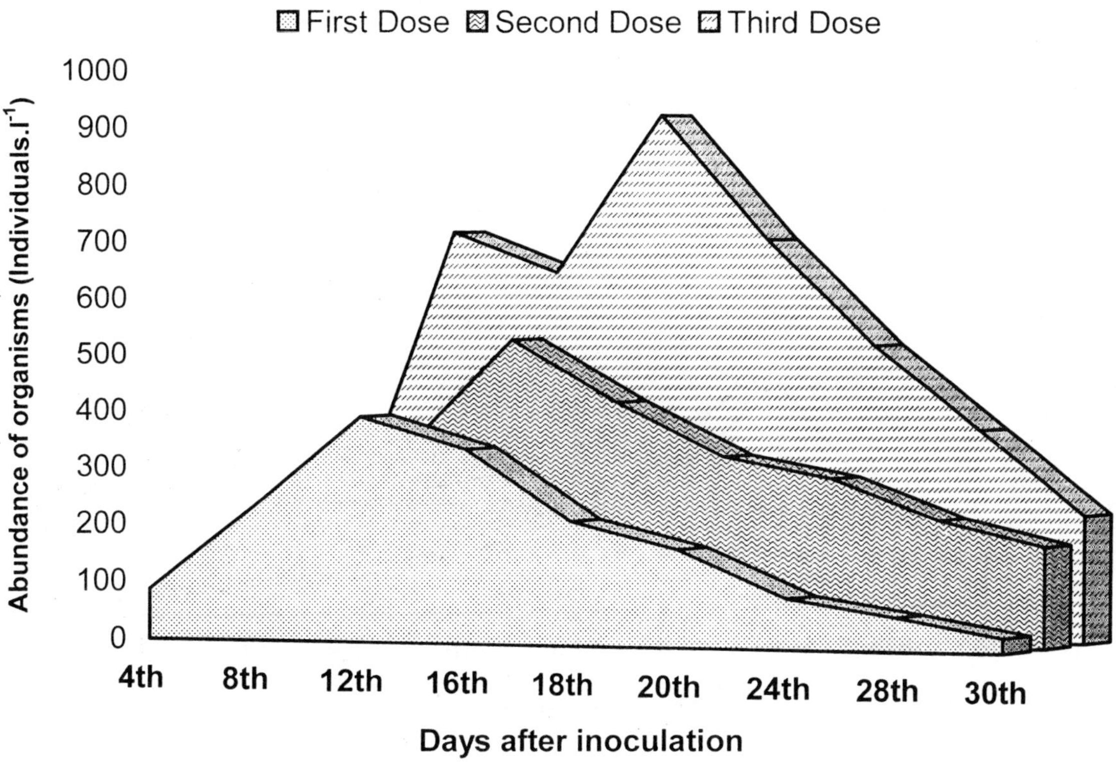

Figure 2.2: Abundance of *Moina micrura* in Three Doses of DWH and PM

Physico-chemical Quality of Water

It was observed that in all the three doses organic manures caused alterations in the physico-chemical parameters (Table 2.1). There was no difference in water temperature among treatments. During entire period of investigation, water temperature in all the three doses varied from 26.1 to 31.7 ^{0}C. The pH ranged from 8.13 to 9.16 in first dose, 8.02 to 9.83 in second dose and 7.23 to 8.82 in third dose. The amount of DO varied from 4.85 to 11.15 mg.l^{-1} in first dose, 3.21 to 9.45 mg.l^{-1} in second dose and 2.97 to 9.20 mg.l^{-1} in third dose. The value of free CO_2 ranged from 7.26 to 41.38 mg.l^{-1} in three doses throughout the study period. The concentration of Ammonia in three doses ranged from 0.57 to 4.33 mg.l^{-1} in first dose, 0.68 to 6.15 mg.l^{-1} in second dose and 0.89 to 10.32 mg.l^{-1} in third dose. During entire period of study the level of nitrate and phosphate varied from 0.12 to 2.29 mg.l^{-1} and 0.89 to 1.68 mg.l^{-1} respectively in three doses.

Discussion

In the present study, the third dose of DWH and PM was optimum for culture of *D. carinata*, *M. micrura* and *C. cornuta* and their higher number was found on 20th day (765 inds.l^{-1}), 18th day (923 inds.l^{-1}) and 20th day (2151 inds.l^{-1}) of inoculation respectively. The data on quantitative analysis supports the finding of earlier researchers. Sharma *et al.* (2003) reported higher zooplankton population dominated by cladocerans (>900 inds.l^{-1}) by using dried water hyacinth (930 kg.ha^{-1}) and urea

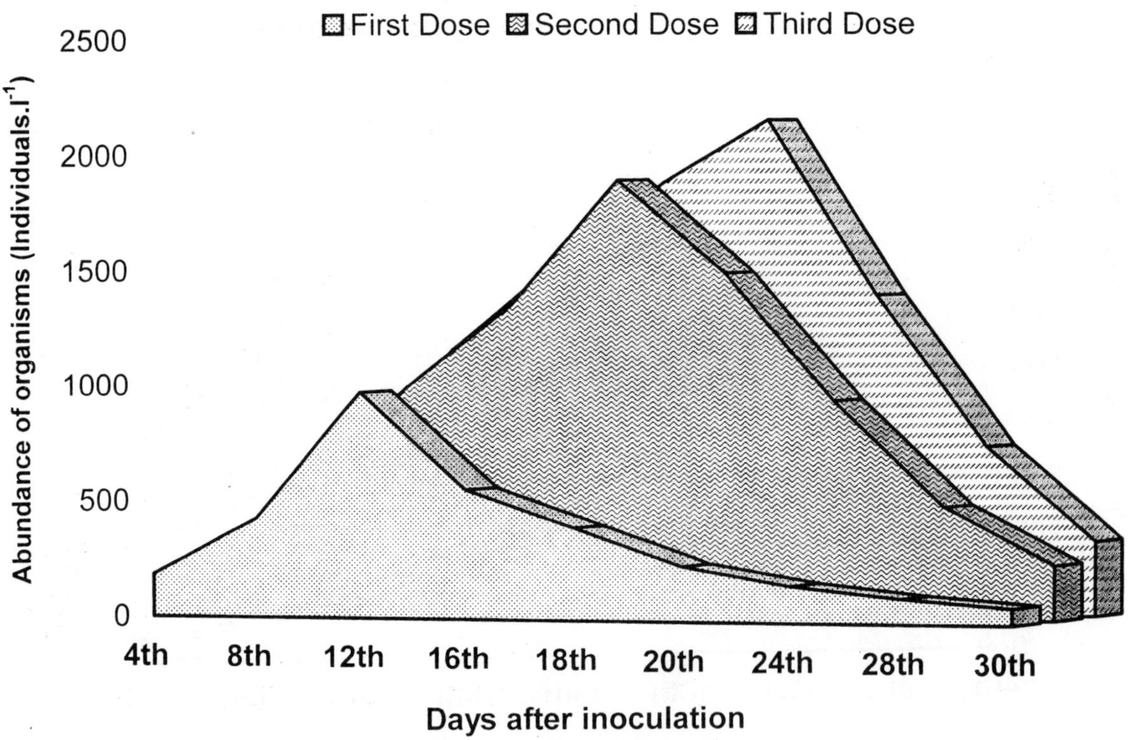

Figure 2.3: Abundance of *Ceriodaphnia cornuta* in Three Doses of DWH and PM

(25 kg.ha^{-1}) in mixed culture system. Srivastava *et al.* (2006) have reported >1900 inds.l^{-1} of C. cornuta by using cow dung, mustard oil cake and poultry manure at higher dose (2.104 kg.m^{-3}). Dhawan and Kaur (2002) have also reported higher zooplankton density (>2000 inds.l^{-1}) from pig dung applied at higher concentration (36 t.ha^{-1}) in polyculture system.

Table 2.1: Physico-chemical Parameters in Three Different Treatments of *D. carinata, M. micrura* and *C. cornuta*

Physico-chemical Parameters	*Organisms*	*First Dose (DWH:PM @ 1.26 g.l^{-1})*	*Second Dose (DWH:PM @ 2.52 g.l^{-1})*	*Third Dose (DWH:PM @ 5.04 g.l^{-1})*
Temp (°C)	*D. carinata*	26.5–31.3	26.2–31.7	26.1–31.5
	M. micrura	26.5–31.5	26.1–31.5	26.3–31.7
	C. cornuta	26.7–31.4	26.3–31.5	26.5–31.7
pH	*D. carinata*	8.13–9.16	8.36–9.83	7.79–8.82
	M. micrura	8.46–9.07	8.02–9.16	7.23–8.76
	C. cornuta	8.25–9.03	8.37–9.54	7.37–8.39
DO (mg.l^{-1})	*D. carinata*	4.85–10.90	4.18–9.45	2.97–8.77
	M. micrura	5.07–11.15	3.21–8.97	3.43–8.33
	C. cornuta	5.58–10.77	4.32–9.23	3.57–9.20

Contd...

Table 2.1–Contd...

Physico-chemical Parameters	Organisms	First Dose (DWH:PM @ 1.26 g.l^{-1})	Second Dose (DWH:PM @ 2.52 g.l^{-1})	Third Dose (DWH:PM @ 5.04 g.l^{-1})
F CO_2 (mg.l^{-1})	*D. carinata*	8.42–17.13	18.26–31.53	14.27–32.17
	M. micrura	7.78–16.65	12.24–29.87	17.53–41.38
	C. cornuta	7.26–17.11	15.67–27.16	16.96–39.87
Ammonia (mg.l^{-1})	*D. carinata*	0.57–3.98	0.81–6.15	0.95–10.32
	M. micrura	0.61–4.33	0.68–4.73	0.97–9.76
	C. cornuta	0.63–4.18	0.73–5.92	0.89–9.83
Nitrate (mg.l^{-1})	*D. carinata*	0.12–0.47	0.36–1.07	0.52–2.13
	M. micrura	0.22–0.53	0.28–0.98	0.53–1.91
	C. cornuta	0.17–0.65	0.36–1.03	0.55–2.29
Phosphate (mg.l^{-1})	*D. carinata*	0.03–0.73	0.07–1.26	0.08–1.98
	M. micrura	0.05–0.42	0.04–1.32	0.08–2.56
	C. cornuta	0.04–0.87	0.06–0.97	0.09–2.17

In the present study the physico-chemical parameters in all the three treatments remained within the favorable range (Delbaere and Dhert, 1996). Organic manures were allowed to decompose for 10 days before inoculation of zooplankton organisms. At the initial days of manuring, the decomposition of DWH and PM resulted in lower values of pH (7.23) and DO (2.97 mg.l^{-1}) but higher values of free CO_2 (41.38 mg.l^{-1}), ammonia (10.32 mg.l^{-1}), nitrate (2.29 mg.l^{-1}) and phosphate (2.56 mg.l^{-1}). Peak population of zooplankton organisms resulted when values of pH, DO, free CO_2, ammonia, nitrate and phosphate ranged between 8.07-8.36, 4.03-5.17 mg.l^{-1}, 9.67-23.15 mg.l^{-1}, 0.68-1.27 mg.l^{-1}, 0.63-0.94 mg.l^{-1} and 0.35-0.87 respectively which was optimum values for cladoceran culture. Manure decay required more dissolved oxygen, which resulted in lower value of dissolved oxygen in culture system. Direct pond fertilization using excessive manure can lead to water quality deterioration including severe depletion of dissolved oxygen (Boyd, 1982). Morris and Mischke (1999) reported that organic manure causes dissolved oxygen problems during initial decomposition. In present study, dissolved oxygen gradually increased with progress of culture duration. Biological oxidation converts the organic matter in manure to carbon dioxide. The higher concentration of free carbon dioxide in culture system combined with water to produce carbonic acid, which caused lowering of pH. At the initial days of manuring pH also lowered due to higher level of ammonia, which on nitrification produced acidity. When manure decay, ammonia is released in culture system (Boyd and Tucker, 1998). In the present study ammonia, was significantly ($P<0.01$) higher in third dose than others. In first, second and third doses, peak population obtained when ammonia concentration ranged between 0.68-0.95 mg.l^{-1}, 0.78-1.06 mg.l^{-1} and 0.96-1.27 mg.l^{-1} respectively. Mitra and Banerjee (1976) reported higher organic carbon, available phosphorus and available nitrogen in aquatic weed water hyacinth.

Conclusion

The result of present study with DWH and PM at the dose 5.04 g.l^{-1} would be more effective in nutrient release and production of cladoceran. In the present scenario these two organic manures would be very economical and helpful for fish farmer, especially use of water hyacinth could be effective in eradicating this weed as well.

Acknowledgements

We are grateful to the Principal, S. M. M. Town Post Graduate College, Ballia for kindly providing the facilities for performing these experiments.

References

Alikunhi, K.H., Chaudhari, H. and Ramachandram, V., 1955. On the mortality of carp fry in nursery ponds and the role of plankton in their survival and growth. *Ind. J. Fish.*, 2: 257–313.

APHA, 1995. *Standard Methods for the Examination of Water and Wastewater*, 19th Edition, APHA, Washington DC.

Arun Kumar, J., 2005. Production of *Spirulina* through application of organic manure. *Fishing Chimes*, 25(7): 48–49.

Boyd, C.E., 1982. *Water Quality Management for Pond Fish Culture*. Elsevier Scientific Publishing Company, New York.

Boyd, C.E. and Tucker, C.S., 1998. *Pond Aquaculture Water Quality Management*. Kluwer Academic Publishers, London, 700 pp.

De Pauw, N., Laurey, P. and Morales, J., 1981.Mass cultivation of *Daphnia magna* (Straus) on rice bran. *Aquaculture*, 25: 141–152.

Delbaere, D. and Dhert, P., 1996. Zooplankton. In: *Manual on the Production and Use of Live Food for Aquaculture*, (Eds.) P. Lavens and P. Sorgeloos, pp. 361–370.

Dhawan, A. and Kaur, S., 2002. Pig dung as pond manure: Effect on water quality, pond productivity and growth of carps in polyculture system. *Naga, The ICLARM Quarterly*, 25(1): 11–14.

Fernando, C.H., 1994. Zooplankton, fish and fisheries in tropical fresh waters. *Hydrobiol.*, 272: 105–123.

Jana, B.B. and Chakrabarti, R., 1993.Life table responses of zooplankton (*Moina micrura* Kurz and *Daphnia carinata* King) to manure application in a culture system. *Aquaculture*, 117: 273–285.

Keke, R., Ofojekwu, P.C., Asala, G.N. and Anosita, J.C., 1994. The effect of partial substitution of groundnut cake by water hyacinth (*Eichhornia crassipes*). *Acta Hydrobiol.*, 32(2): 235–244.

Kumar, S., Sharma, J.G. and Chakrabarti, R., 2000. Quantitative estimation of proteolytic enzyme and ultrastructure study of anterior part of intestine of Indian major carp (*Catla catla)* larvae during ontogenesis. *Curr. Sci.*, 79: 1007–1011.

Kumar, S., Srivastava, A. and Chakrabarti, R., 2005. Study of digestive proteinases and proteinase inhibitors of *Daphnia carinata. Aquaculture*, 243: 367–372.

Lee, H.L., 1982. Preliminary studies on the culture on *Moina* sp. using organic waste. *Hons. Thesis*, University of Singapore, pp. 107.

Mitra, E. and Banerjee, A.C., 1976. Utilization of higher aquatic plants in fishery water. In: *Aquatic Weed in S.E. Asia*, (Eds.) C.K. Varshney and J. Rzoska. Dr. W. Junk B.V. Publishers, The Hague, p. 375–381.

Morris, J.E. and Mischke, C.C., 1999. Plankton management of fish culture ponds. Technical Bulletin Series, Vol. 114. Iowa State University Agricultural Experiment Station, 8 pp.

Sharma, O.P., 2003. Zooplankton production using *Eichhornia cressipes* as biofertilizer. *Fishing Chimes*, 23(7): 42–45.

Sharma, J.G. and Chakrabarti, R., 2000. Replacement of live-food with refrigerated-plankton food for *Cyprinus carpio* (L.) larvae cultured with three different types of biological filters. *Curr. Sci.*, 79: 214–219.

Singh, D.K. and Datta Munshi, J.S., 1991. Some aspects of biology of Cladocerans of River Ganga. *J. Freshwater Biol.*, 3: 187–191.

Srivastava, A., Rathore, R.M. and Chakrabarti, R., 2006. Effects of four different doses of organic manures in the production of *Ceriodaphnia cornuta*. *Bioresource Technology*, 97: 1036–1040.

Tay, S.H., 1980. Experiments on the cultivation of *Moina micrura* Kurz. Pri. Prod. Dept. Singapore (Unpubl.), pp.4.

Chapter 3

Effet of Photosensitizer on Melanophore Responses of Blue Gouramei (*Trichogaster trichopterus*)

☆ *O.A. Asimi, P.P. Srivastava and T.H. Shah*

Introduction

Long before aquariums we've known that fishes change color in response to their background, and that they change colour during exercise and courtship. These changes in appearance are under the control of pigment containing cells called "chromatophores". These are the same types of cells present in some invertebrates: crustaceans, insects, cephalopods, and are found in other cold-blooded vertebrates (reptiles, amphibians). The first to review the literature on the physiology and behavior of animal color (Animal Colour Changes) was Parker (1948). This work was further compiled and added to by Bagnara and Hadley (1973) in the third volume of Fish Physiology: Chromatophores and Pigments. Importantly this last work covers how fishes use their chromatophores in a variety of responses to internal and external stimuli. The ability to change skin colour and pattern is essential for survival in many species. In species such as the flatfish or the chameleon, the change occurs rapidly and are controlled by hormonal or nerve signals. The slowly evolving colour change occurs in response to continuous exposure to stimuli, as can be seen by prolonged exposure of skin to sunlight. A particularly striking example occurs in members of the marine angelfish genus *Pomacanthus*. A number of diseases are attributed to defective pigmentation, including the condition vitiligo which affects approximately 1 per cent of the population. This disease is a result of destruction of melanocytes and causes patches of white skin (M. McClure, 1998). Methoxsalen, which is used in the present investigation, belongs to furocumarin group of organic chemical compounds and is a very potent photosensitizing agent when combined with exposure to sunlight thereby, being very useful in the treatment of Vitiligo.

Ornamental fish are acceptable to consumers if they have striking and vibrant colours; hence people involved in this trade are constantly exploring methods of enhancing skin colouration. Colour enhancement through the use of carotenoids in feed has been confirmed by a number of authors (Fey and Meyers, 1980; Ako, Tamaru, Asano, Yuen and Yamamoto, 2000; Alagappan, Vijula and Sinha, 2004). Sinha and Asimi (2007) reported that the China rose (*Hebiscus rosasinensis*) petal is a potent natural carotenoid source for goldfish to enhance its colour and also accelerate its gonadal development. In the present investigation after seeing the various aspects of methoxsalen, an attempt has been made to study its effect on the body colouration of aquarium fishes. The blue Gourami (*Trichogaster trichopterus*), being an excellent ornamental fish, was chosen for the study. The experiment was conducted with the following objectives:

1. To evaluate the influence of dietary supplemented photosensitizer (methoxsalen) on the melanophores, and
2. To evaluate the persistence of changes in melanophores after discontinuing the photosensitizer supplemented feeds.

The finding will be very useful to understand the changes brought in the behavior of melanophores after feeding with photosensitizer (methoxsalen) through dietary supplementation to render ornamental fishes look paler and/or darker. The aggregation and dispersion phenomenon of chromatophores has direct impact relation with the market value of ornamental fish.

Materials and Methods

Twelve glass aquaria of 40 liter capacity each were properly cleaned and filled with water and maintained at 30 liter throughout the experiment. Aeration was provided round the clock. Aquaria were divided into four groups of triplicates. The experimental fish, blue Gourami (*Trichogaster trichopterus*) was procured from local aquarium shop and stocked @10 fish/aquarium after 15 days of acclimatization. The experiment was conducted over a period of 31 days from 1st December, 2001 to 1st January, 2002. A basal feed was prepared using Fish meal, Acetes, Soybean meal, Groundnut oil cake, Wheat flour, Vitamin + Mineral mixture and Vegetable oil. This basal feed was then divided into four equal portions and methoxsalen, in different concentration, was added to three portions. Thus in feed E_1, 20 mg methoxsalen was added per kg of basal feed, in feed E_2, 100 mg methoxsalen and in feed E_3, 500 mg methoxsalen was added per kg of basal feed. After mixing thoroughly required amount of water was added separately in E_1, E_2 and E_3. The fourth portion of the basal feed was used as control feed. For initial 15 days the experimental and control feed was fed after which the scale samples were collected from lateral side of the fish. All the fish were fed with control diet for another 15 days to see the sustainability of the pigmentation due to action of Methoxsalen.

Table 3.1: Details of Feed Developed in the Laboratory

	Control	E_1	E_2	E_3
Basal feed Ingredients	Fish meal (40 g), Acetes (20 g), Soybean meal (10 g), Groundnut oil cake (10 g), Wheat flour (10 g), Vitamin + Mineral mixture (4 g) and Vegetable oil (6g)	Fish meal (40g), Acetes (20 g), Soybean meal (10 g), Groundnut oil cake (10 g), Wheat flour (10 g), Vitamin + Mineral mixture (4 g) and Vegetable oil (6g)	Fish meal (40g), Acetes (20 g), Soybean meal (10 g), Groundnut oil cake (10 g), Wheat flour (10 g), Vitamin + Mineral mixture (4 g) and Vegetable oil (6g)	Fish meal (40g), Acetes (20 g), Soybean meal (10 g), Groundnut oil cake (10 g), Wheat flour (10 g), Vitamin + Mineral mixture (4 g) and Vegetable oil (6g)
Photosensitizer (Methoxsalen)	Nil	20 mg kg^{-1} (20 ppm)	100 mg kg^{-1} (100 ppm)	500 mg kg^{-1} (500 ppm)

The fish were fed *ad libitum* at 11:00 am everyday and one hour after feeding, they were exposed to sunlight for half an hour (12.00–12.30 pm). Exposure to sunlight was to let the photosensitizer work inside the body of the fish. The above procedure was continued for 15 days after which the scales were taken from lateral side of the fish and preserved in 6 per cent formalin. The experimental diets were stopped and fish were fed only control diet for another 15 days to see the persistence of the photosensitizer. The scales of all the fish i.e, control and experimental diet fed fish were taken once again and fixed on glass slides to be viewed under compound microscope for chromatophores counts. Water quality parameters were also analyzed using standard methods (APHA, AWWA and WEF 1998).

Results

The scale of the fish fed on control feed showed no darkening other than normal, either on the centre or scale ridges (Figure 3.1) whereas the fish fed on experimental feed, E_1 (methoxsalen @20 ppm in the diet) showed a marked darkening of the scale (Figure 3.2). An increase in the darkening of the scale as well as the number of pigment granules was observed in the fish fed on E_2 feed (methoxsalen @100 ppm) as can be seen in Figures 3.3 and 3.4. This increased number of pigments/melanophores is clearly seen at higher magnification (250 X) (Figure 3.4). The fish fed with experimental diet E_3 (methoxsalen @500 ppm) showed further darkening of the scales and increased number of pigments/chromatophores (Figure 3.5). At higher, magnification (250 X) the aggregation of melanophore is very well seen (Figure 3.6). Physico-chemical parameters showed acceptable values. The water temperature varied in the range of 21 ± 2°C, while pH was between 7.2–7.6, dissolved oxygen content ranged between 6.6–6.8 ppm, free CO_2 was nil, while as alkalinity ranged between 164–165 ppm.

Discussion

The dispersing action of methoxsalen is very well documented in higher dose (E_3, 500 mg/kg) and this effect is clearly visualized at higher magnification and trend of dispersing action in the scales of *Trichogaster trichopterus* (Photo-6). Similar to the present finding the pigment concentration of *Salmo salar* was observed by Buttle *et al.* (2001) and Storebakken *et al.*(1987), in *Oncorhynchus mykiss* (No and Storebakken, 1991; March and Macmillan, 1996) and *O. kitsutch* (Smith, et al., 1992). Hama and Hasegawal (1967) while studying the chromatophores of *Oryzias latipes* (medaka) reported that pigments in the feed exhibited a profound effect on the behaviour of chromatophores. The findings of Sugimoto *et al.* (2000) on the skin pigmentation of the medaka have full support to the present finding. The cytoskeletal structure of dermal chromatophores in *Oryzias latipes* was studied by Obika and Fukuzawa (1993) and they observed that the microtubule system was most prominent in melanophores and played important role in aggregation and dispersal of chromatophores. Similar reasons of the colour change in the present experiment by using the photosensitizer, methoxsalen. In a recent work, on the use of mammalian endothelins to effectively disperse the light scattered organelles in leucophores in medaka (Fujita and Fujii, 1997) in a dose depended manner have concurrence with the present studies on labyrinth fish blue Gourami (*Trichogaster trichopterus*). Oshima *et al.* (2001) reported that the use of Melanin Concentrating Hormone (MCH) at a low dose (<1mM) induced pigment aggregation and this aggregation state was maintained in the presence of MCH. Aggregation was immediately followed by some re-dispersion, even in the continued presence of MCH, which let to an apparent decrease in aggregation. The mechanism of action of the drug, methoxsalen, used in the present study exhibits almost similar mechanism for colour aggregation and dispersion in blue Gourami (*Trichogaster trichopterus*).

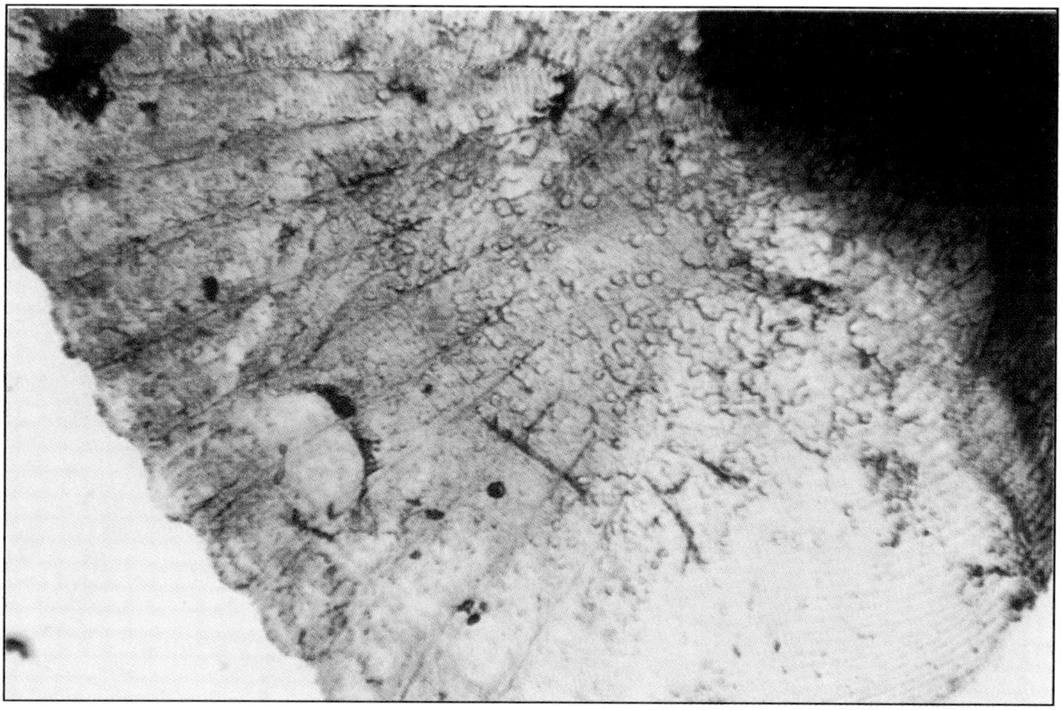

Figure 3.1: Scale of Blue Gourami (*Trichogaster trichopterus*) Fed on Control Feed (100 X)

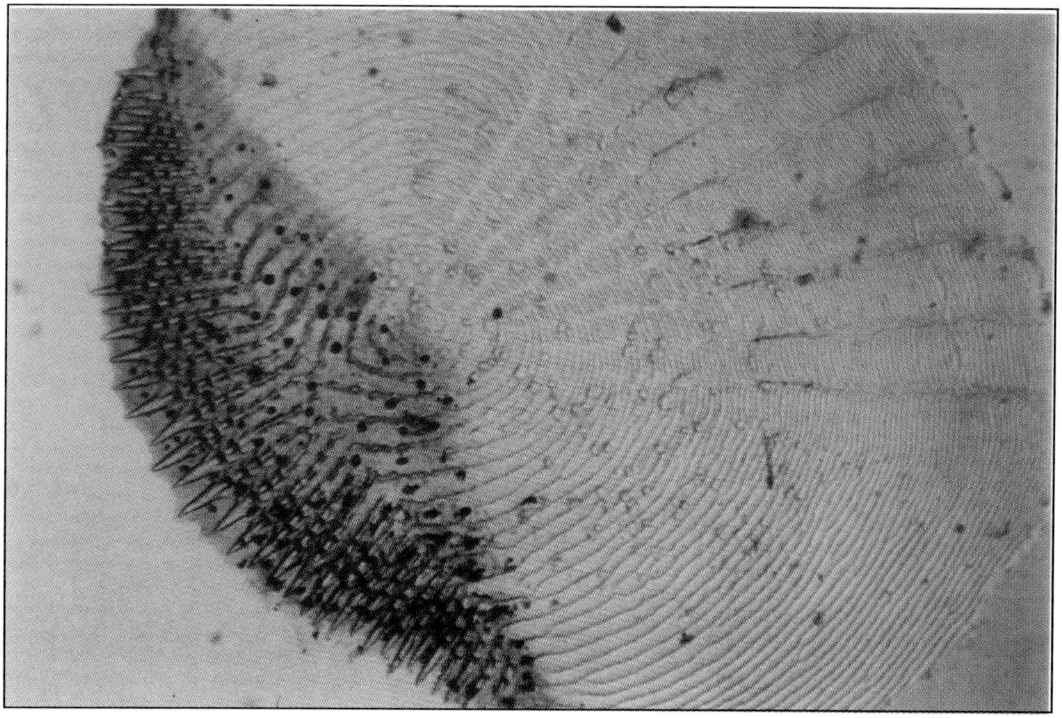

Figure 3.2: Scale of Blue Gourami (*Trichogaster trichopterus*) Fed Diet E1 (100 X)

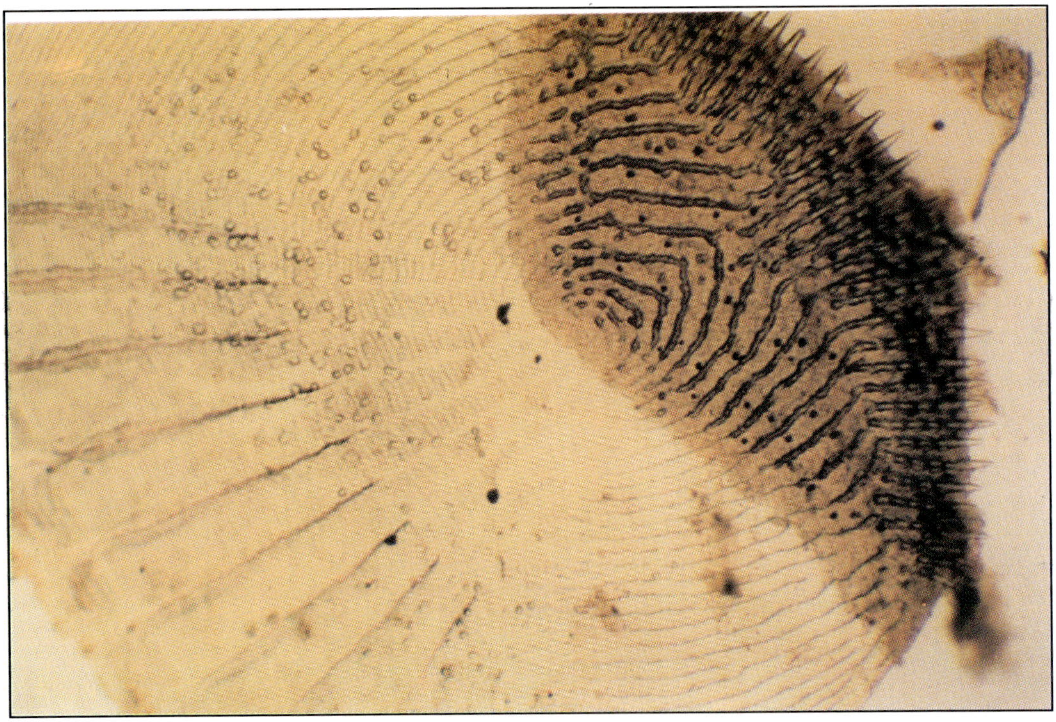

Figure 3.3: Scale of Blue Gourami (*Trichogaster trichopterus*) Fed on Diet E_2 (100 X *trichopterus*)

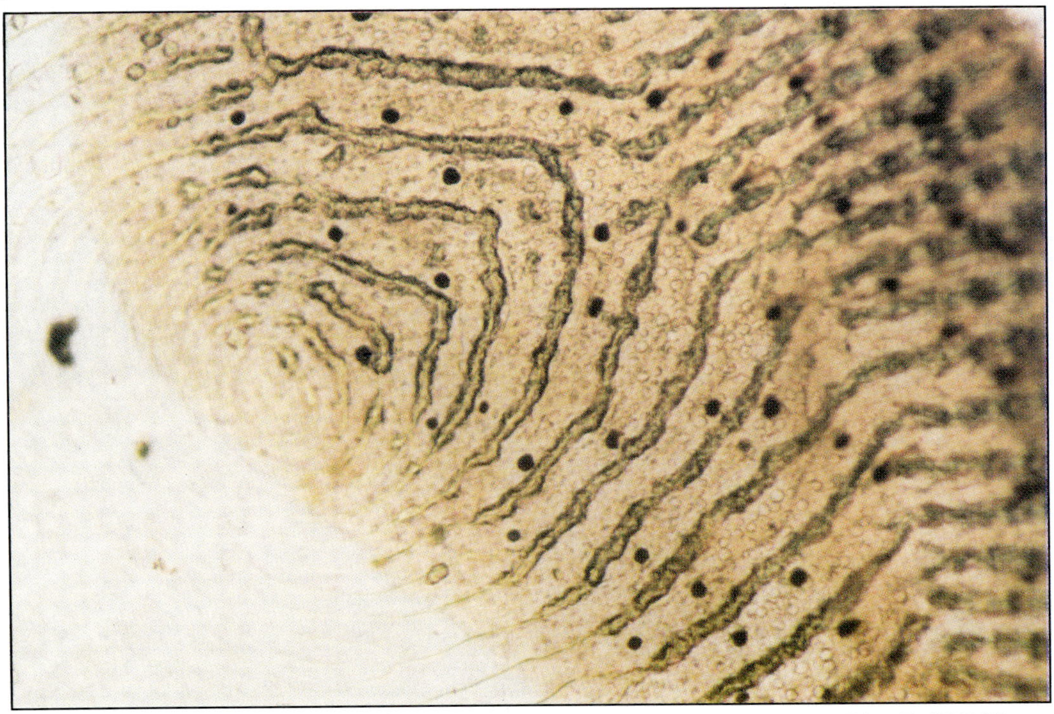

Figure 3.4: Scale of Blue Gourami (*Trichogaster trichopterus*) Fed on Diet E_2 (250 X)

Figure 3.5: Scale of Blue Gourami (*Trichogaster trichopterus*) Heavily Darkened Fed on Diet E_3 (100X)

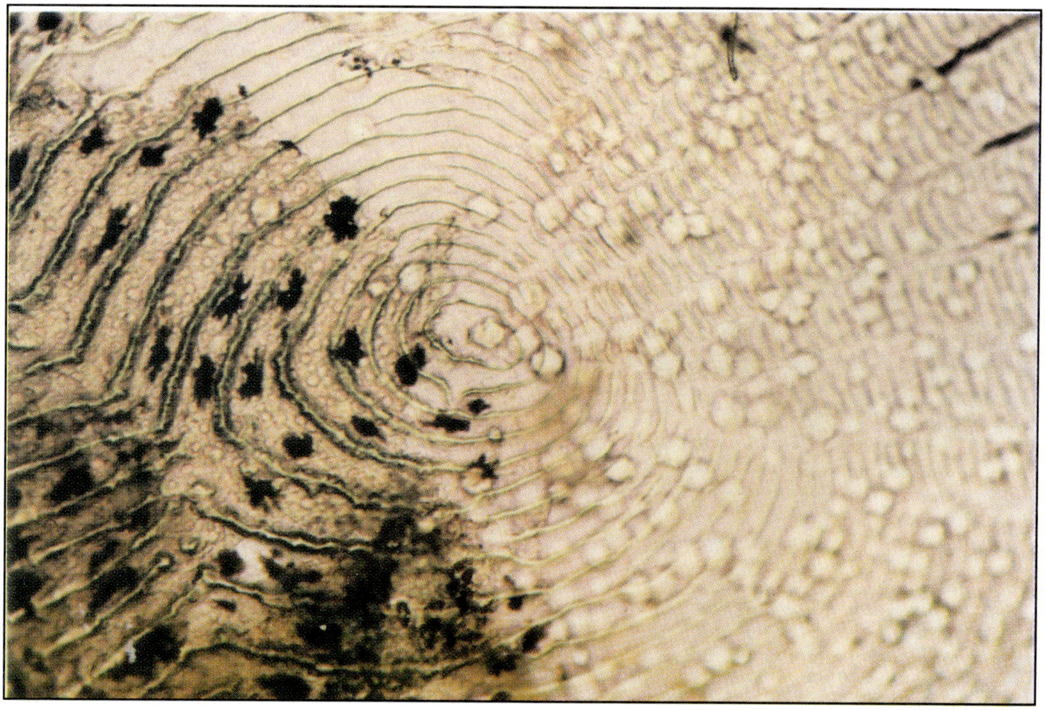

Figure 3.6: Scale of Blue Gourami (*Trichogaster trichopterus* Fed on Diet E_3 (250 X). Increased darkening of the scale as well as aggregation of melanophores is clearly visible.

Conclusion

The findings of the present study are very useful to understand the changes brought in the behaviour of melanosomes in the melanophores on feeding with photosensitizer methoxsalen, through dietary supplementation, to render ornamental fishes look paler and/or darker. However, some more studies are needed to be conducted to evaluate the physiological stresses and biochemical changes, if any, that may come about in the fish, due to the addition of methoxsalen before taking any commercial advantage out of it.

References

Ako, H., Tamaru, C.S., Asano, L., Yuen, B. and Yamamoto, M., 2000. Achieving natural colouration in fish under culture. *UJNR Technical Report*, 28.

Alagappan, M., Vijula, K. and Sinha, A., 2000. Utilization of spirulina algae as a source of carotenoid pigment for blue gouramis (*Trichogaster trichopterus* Pallas). *Journal of Aquariculture and Aquatic Sciences*, 10: 1–11.

APHA, AWWA, WEF, 1998. *Standard Methods for the Examination of Water and Wastewater*, 20th edn. (Eds.) L.S. Clesceri, A.E. Greenberg and A.D.Eaton. American Public Health Association, Washington, DC, USA, 1060 pp.

Bagnara, J. and Mac Hadley, 1973. Chromatophores and pigments. In: *Fish Physiology, Vol. 3,* (Eds.) W.S. Hoar and Randall. Academic Press, New York, London.

Fey, M. and Meyers, S.P., 1980. Evaluation of carotenoid-fortified flake diets with the pearl Gourami, *Trichogaster leeri. Journal of Aquariculture*, 1: 15–19.

Fujita, T. and Fijii, R., 1997. Endothelins disperse light-scattering organelles in leucophores of the medada, *Oryzias latipes. Dep. Biomol. Sci. Fac. Sci.*, Toho Univ., Miyonna, Funabashi, Chiba 274, Japan.

March, B.E. and Macmillan, C., 1996. Muscle pigmentation and plasma concentrations of astaxanthin. *Progressive Fish Culturist*, 58: 178–186.

McClure, M., 1998.Growth, Shape Change, and the development of pigment patterns in fishes of the genus Danio *(Teleostei: Cyprinidae). Ph.D. Thesis*, Cornell University.

No, H.K. and Storebakken, T., 1991. Pigmentation of rainbow trout with astaxanthin and canthaxanthin in freshwater and seawater. *Aquaculture*, 101: 123–134.

Obika, M. and Fukuzawa, 1993. Cytoskeletal architecture of dermal chromatophores of the freshwater teleost, *Oryzias latipes. Pigment Cell Res.*, 6(6): 417–422.

Oshima, N., Nakamaru, N., Araki, S. and Sugimoto, M., 2001.Comparative analysis of the pigment–aggregating and dispersing actions of MCH on fish chromatophores. *Comp. Biochem. Physiol. C Toxicol. Pharmacol.*, 129(2): 75–84.

Parker, G.H., 1948. *Animal Colour Changes and their Neurohumours: A Survey of Investigations* 1910–1943. Cambridge University Press.

Sinha, A. and Oyas Amed Asimi, 2007. China rose (*Hibiscus rosasinensis*) petals: A potent natural carotenoid source for goldfish (*Carassius auratus* L). *Aquaculture Research*, 38: 1123–1128.

Smith, B.E., Hardy, R.W. and Torrissen, O.J., 1992. Synthetic astaxanthin deposition in pan-size Coho salmon (*Onchorhynchus kitsutch*). *Aquaculture,* 104: 105–119.

Storebakken, T., Foss, P., Schiedt, K., Austreng, E., Liaaen-Jensen, S. and Manz, U., 1987. Carotenoids in diets for salmonids IV. Pigmentation of Atlantic salmon with astaxanthin, astaxanthin di-palmitate and canthaxanthin. *Aquaculture,* 65: 279–292.

Sugimoto, M., Uchidea, N. and Hatayama, M., 2000. Apoptosis in skin pigment cells of the medaka, *Oryzias latipes* (Teleostei), during long-term chromatic adaptation. *Cell Tissue Res.,* 301(2): 205–216.

Chapter 4

Dietary Effect of Chitin, Chitosan and Levamisole on Survival, Growth and Production of *Cyprinus carpio* Against the Challenge of *Aeromonas hydrophila* Under Temperate Climatic Conditions of Kashmir Valley

☆ *Sajid Maqsood, Samoon, M.H. and Prabjeet Singh*

Introduction

Aquaculture is a fast developing industry. However, unmanaged fish culture practices and adverse environmental conditions affect the fish health leading to production losses. Thus, fish farmers have to carryout careful husbandry practices (Sakai, 1999). In order to achieve optimal fish production, better prophylactic, diagnostic and therapeutic measures are warranted during fish farming operations. Use of expensive chemotherapeutants and antibiotics for controlling disease have widely been criticized for their negative impact like residual accumulation in the tissue, development of the drug resistance and immunosuppression, thus resulting in reduced consumer preference for food fish treated with antibiotics (Anderson, 1992). Hence, instead of chemotherapeutic agents, increasing attention is being paid to the use of immunostimulants for disease control measures and growth increment in aquaculture.

Fishes are usually reared in holding spaces such as ponds, tanks or cages and efforts have been made to increase productivity per unit area. High density or intensive fish culture system tends to adversely affect the health of reared fish and this condition tends to produce a poor physiological environment for fish and increases their susceptibility to infection (Kim *et al.*, 2003). Drug resistance and poor growth in farmed fish is a major constraint in aquaculture industry. Growth and immune stimulants holds the promise to improve fish health and subsequently control fish diseases in aquaculture. The use of immunostimulants for the prevention of fish diseases is considered an attractive and promising area (Anderson, 1992; Secombes, 1994). Immunostimulants are valuable for the prevention and control of fish diseases in aquaculture as these represent an alternative and supplementary treatment to vaccination. The immunostimulants also have additional advantages, such as growth enhancement and increase in the survival rates of the fishes under stress (Heo-GangJoon *et al.*, 2004). The development of strategies to control the pathogen load and immunoprophylactic measures is required to improve the overall fish production. The use of immunostimulants as dietary supplement can improve the innate defense mechanism of fish providing resistance to pathogens during the period of high stress. Hence, the use of immunostimulants is being introduced in the routine fish farming operations as prophylactic measure. Unlike antibiotics and live vaccines reported to have negative effects on the consumer and environment, there are no reports of any such adverse effect through the use of immunostimulants in fish farming (Anderson, 1992; Secombes, 1994). The mode of administration of immunostimulants is very important in aquaculture and should be acceptable with regard to labour input, convenience of the method and stress state of the fish.

Chitin is an insoluble, linear β-1,4-linked polymer of N-acetyl-D-glucosamine, one of the most abundant polysaccharide in nature and a common constituent of insect and crustacean exoskeleton and fungal cell wall. Chitin is commercially manufactured from the shrimp and crab shells. Chitin is non-toxic biodegradable and biocompatible, for which reason chitin and its derivatives have been used in medical practice (Shibata *et al.*, 1997a). Increased protection against *Aeromonas salmonicida* has been observed in brook trout when injected with chitin. Injection of abalone extract and chitin increase phagocytic response and natural killer cell activity in fish (Stickney, 2000). Further, chitin is reported to provoke the immunostimulation in a very short time, thus making it an interesting candidate for incorporation in fish diet formulation (Esteban *et al.*, 2000a).

Chitosan is a linear homopolymer of β-(1, 4)-2-amino-deoxy-D-glucose and is prepared by the alkaline deacetylation of chitin obtained from crab shell. Chitosan is used as an immunostimulant in aquaculture to protect salmonids and carps against bacterial diseases (Anderson and Siwicki, 1994a; Siwicki *et al.*, 1994). Dietary chitosan has a positive effect on growth of common carp which ultimately leads to increase in the overall production of fish under pond condition (Gopalakannan and Venkatesan, 2006)

Levamisole, a synthetic phenylimidazolthiazole has been extensively used in both humans and veterinary medicine as an anti-helminthic agent (Janssen, 1976). The immunostimulatory and growth enhancing potential of levamisole in fish is of considerable interest in the present scenario of aquaculture in combating the bacterial and parasitic disease of fish because the U.S Food and Drug Administration (FDA) have approved it for treatment of helminth infections in ruminants. Levamisole, which is a synthetic phenylimidathiazole, has been shown to have the ability to up-regulate nonspecific immune response as well as growth in carps, rainbow trout and gilthead sea bream (Stickney, 2000).

Common carp is an important candidate species for aquaculture in Kashmir valley where carp farming has started picking up pace during the last one decade. *Aeromonas hydrophila* has been reported to cause fin rot disease in hatchery reared *Cyprinus carpio* under temperate climatic conditions of Kashmir valley leading to production losses (Hussain *et al.*, 2005). The authors recommended use of multiple antibiotics as a bath for controlling fin rot disease. However, as mentioned above, controlling diseases with antibiotics is not a safe procedure in aquaculture practices. Hence, the present study was conducted using immunostimulants for enhancing growth and survival of common carp under pond condition in temperate climatic condition of Kashmir valley.

Materials and Methods

Fish and Rearing Conditions

Healthy and disease free advanced fingerlings of common carp (*Cyprinus carpio*) having an average weight of 16± 2g and total length of 11± 2cms were procured from Fish Farm of Faculty of Fisheries. The experimental stock was acclimatized for a period of 2 weeks in concrete ponds containing same source of water which was used for conducting the experimental trial and the stock was fed on control diet (D_1). Four experimental concrete ponds with proper inlets and outlets and measuring 6.0 x 6.0 x 1.5 m were used for conducting the experiment. These were thoroughly treated with quick lime and disinfected with $KMnO_4$. The ponds were then filled with the spring water and a uniform water column of 1.2 m was maintained through out the experimental period. Pond water replenishment was carried out every week by replacing 40-50 per cent of the pond water. After acclimatization, the fish were divided into four groups of 60 specimens each and were stocked in the experimental ponds. Water quality parameters like temperature, dissolved oxygen, pH and free CO_2 were recorded on weekly basis. Water and air temperature was recorded by a standard quality thermometer. Dissolved oxygen and pH were recorded with digital DO and pH meter respectively. Free CO_2 was determined titrematically by following the standard procedures (A.P.H.A. 1998). The source of water used for conducting the experimental trial was natural spring water.

Experimental Feed

Feed ingredients viz, groundnut oil cake, rice bran, soybean meal, fish meal, and wheat flour were procured, screened and subjected to proximate analysis following standard procedure (A.O.A.C. 2006). All the ingredients were properly weighed as per their inclusion rates in the four experimental diets (Table 4.1). All the weighed ingredients were grounded separately in an electric grinder and thoroughly mixed and water added in sufficient quantity and the whole mixture was steam cooked for 20-25 minutes. Later the required quantities of immunostimulants along with vitamin and mineral mixture were incorporated and mixed thoroughly and evenly to the prepared dough. The resultant dough of desired consistency was passed through a hand pelletizer with a die of 3mm and the pellets were air dried under shade. The dried pelleted diets were packed in air tight polythene bags and stored at -20°C. Diet D_1 served as control diet as it was not supplemented with any immunostimulant, where as diets D_2, D_3 and D_4 comprised of same ingredients as that of D_1 but these were supplemented with Chitin (1 per cent), Chitosan (1 per cent) and Levamisole (250mg/kg of diet) respectively. Chitin was procured from Hi-media (India), chitosan and levamisole were procured from Sigma (USA).

Experimental stock in all the treatments was fed twice daily for a period of 90 days. The feeding rate was @ 5 per cent of their body weight. The feed was offered in the feeding trays, which were immersed in the ponds at a depth of 0. 7 m and inspected to monitor the consumption of feed which was always found to be consumed in full with in an hour.

Bacterial Strain and Challenge Study

A virulent strain of *Aeromonas hydrophila* received in Tryptose Soya Agar Slants (TSA) from IMTECH (Institute of Microbial Technology), Chandigarh was maintained at 4°C in the Division of Veterinary

Microbiology and Immunology, SKUAST-K, Shuhama. From this slant culture, sub-cultures were maintained on Tryptose Soya Agar (TSA) slants (Hi-media, Mumbai) at 5°C. A stock culture in Tryptose Soya Broth (TSB) (Hi-media, Mumbai) was maintained at -40°C with 0.85 per cent NaCl (w/v) and 20 per cent (v/v) glycerol to provide stable inocula throughout the study period as followed by Chabot and Thunne, 1991; Yadav *et al.*, 1992.

The fingerlings in all the groups were challenged with 100 μl of *Aeromonas hydrophila* at a concentration of $1.5\pm0.3 \times 10^6$ CFU ml^{-1} in PBS as a medium. The bacterial suspension in PBS was inoculated intra-peritoneally in all specimens of all the groups by 1ml insulin syringe on 30th day and the specimens were re-challenged on 58th day. Due care was taken to avoid any injury while challenging the experimental stock with *A.hydrophila*. All the challenged specimens were released back into their respective ponds and were observed for their response against the injected bacterial strain.

Table 4.1: Ingredient Composition of Experimental Diets

Ingredients	*Inclusion Rate (Per cent)*			
	D_1 (Control)	*D_2*	*D_3*	*D_4*
GOC	32.00	32.00	32.00	32.00
Rice bran	26.10	26.10	26.10	26.10
Wheat flour	20.05	20.05	20.05	20.05
Soybean meal	15.90	15.90	15.90	15.90
Fish meal	03.95	03.95	03.95	03.95
Vitamin and mineral mixture	02.00	02.00	02.00	02.00
Chitin	–	1 per cent	–	–
Chitosan	–	–	1 per cent	–
Levamisole	–	–	–	250mgkg^{-1} diet

Experimental Regime

The stock was released @ 60 specimens each in experimental ponds P_1, P_2, P_3 and P_4 and fed on diets D_1, D_2, D_3 and D4 respectively.

Sampling Schedule

30th day ⟶ 1st infection with *Aeromonas hydrophila.* (In all the groups)

58th day ⟶ 2nd infection with *A. hydrophila.* (In all the groups)

Length and weight of 10 randomly selected specimens from each treatment were recorded at fortnightly intervals and ration was adjusted accordingly on the basis of fish biomass. The fortnightly recorded data was used for calculating the Feed Conversion Ratio (FCR), Specific Growth Rate (SGR) and net fish production.

Mortality was recorded through out the period of study and Relative Percentage Survival (RPS) was calculated as per the Baulny *et al.*, 1996.

Growth Parameters

The recorded data on weight was used for calculation of Feed Conversion Ratio (FCR) and Specific Growth Rate (SGR). On each sampling day, the SGR or per cent body weight increase per day and FCR for all the experimental groups was calculated according to Ricker (1979) as follows:-

$$SGR = \frac{\text{Ln of Final weight} - \text{Ln of Initial weight}}{\text{t (time interval in days)}} \times 100$$

$$FCR = \frac{\text{Feed given (dry weight)}}{\text{Weight Gain (wet weight)}}$$

Mortality

Recorded mortality data was used for calculating Relative Percentage Survival (RPS) following Amend (1981)

$$RPS = 1 \frac{\text{[Mortality (per cent) in treated group]}}{\text{[Mortality (per cent) in control group]}} x100$$

Statistical Analysis of the Experimental Data

The experimental data was subjected to the statistical analysis following the Completely Randomized Design (CRD) and the statistical difference between the treatment means and within the treatment means was assessed by two way analysis of variance (ANOVA) techniques followed by Duncan's multiple range test using statistical package (SPSS) to find out the significant difference at 5 per cent level ($P<0.05$) of significance.

Results and Discussion

Water Quality Parameters

The dissolved oxygen (DO) content of water throughout the experimental period ranged between 7-10 mg/l, pH ranged between 7–8.5, temperature of the water in the ponds ranged between 31°C during August and 19°C during the end of October. The Free CO_2 content of the pond water ranges between 0 to 5 mg/l.

SGR and FCR

The FCR values showed the decreasing trend up to 45th day in the chitosan and levamisole supplemented group and increasing trend in control (infected) and chitin supplemented group through out the experimental trial. The highest value (2.4) was recorded in chitin supplemented group and lowest (1.71) in chitosan supplemented group. There was a significant ($p<0.05$) difference in FCR values between control (infected) and chitosan and levamisole supplemented group but FCR values of control (infected) and chitin supplemented groups do not differ significantly ($p>0.05$).

The mean SGR values of control (infected) group differs significantly ($p<0.05$) from chitosan and levamisole supplemented group but the mean SGR value of control (infected) group do not differ significantly ($p>0.05$) from chitin supplemented group through out the experimental trial. The mean SGR values of chitosan and levamisole supplemented groups differ significantly ($p<0.05$) up to the 45th day, after which the value do not differ significantly ($p>0.05$). The highest SGR values were recorded in chitosan supplemented group and lowest in chitin supplemented group.

The results of the present study shows that dietary chitosan and levamisole supplementation enhances the SGR of *Cyprinus carpio*, where as chitin supplementation depressed the growth below that of fish fed with control diet. Similar findings were reported by Gopalakannan and Venkatesan (2006). Present findings contradict with the results of Kono *et al.* (1987) and Shiau and Yu (1999). Feeding of supplemented diet containing 10 per cent chitin, chitosan or cellulose did not affect the growth of red sea bream, Japanese eel and yellow tail (Kono *et al.*, 1987). On the contrary, Shiau and Yu (1999) observed depressed growth in tilapia after feeding chitin and chitosan at 2 per cent, 5 per cent and 10 per cent level. They also speculated that the depressed growth in tilapia may be due to interference of chitosan and chitin in the absorption of nutrients. However, in the present study fishes were fed with chitin (1 per cent), chitosan (1 per cent) and levamisole (250mg kg^{-1}) supplemented diet which are lower than the dosage or levels used by Kono *et al.* (1987) and Shiau and Yu (1999). Chitosan may play a crucial role in enhancing the digestion and absorption of nutrients when incorporated at lower levels. However, fish fed with chitin showed depressed growth after 30 days of feeding. This may be due to the fish developing tolerance to chitin and continuous feeding of chitin supplemented diet may have created stress to the fish. Depressed growth by chitin supplementation is also reported by Gopalakannan and Venkatesan (2006). Esteban *et al.* (2000b) demonstrated that chitin supplementation do not have any significant effect on the growth of fish.

Siwicki and Korwin-Kossakowski (1988) proved that levamisole stimulates the growth of *Cyprinus carpio* larvae without affecting the survival and rate of development. The fishes treated with levamisole were larger and heavier by the end of experimental study when compared to the control (Mulero *et al.*, 1998). Precedent exists for growth enhancing effect of levamisole as observed in carp fingerlings (Siwicki and Korwin-Kossakowski, 1998).

In the present investigation, chitosan and levamisole show positive effect on growth of common carp, depicting significantly higher SGR values and lower FCR values when compared to the control (infected) group. Alvarez Pellitero *et al.*, 2006 also reported that specific growth rate was higher in the levamisole 500mg Kg^{-1} treated groups than in the control (C). Niki *et al.* (1991) also found that immunostimulant (levamisole) significantly increases the SGR in fish when compared with the control group. Misra *et al.* (2005) reported that feeding rohu with 250 and 500 mg of β-glucan kg^{-1} diet resulted in significantly ($p<0.05$) higher SGR and lower FCR values than shown by control fish.

Table 4.2: Feed Conversion Ratio (FCR) (Mean ± SE)

Days/ Treatments	*Day 15 (15th Aug)*	*Day 30 (30th Aug)*	*Day 45 (14th Sep)*	*Day 60 (29th Sep)*	*Day 75 (14th Oct)*	*Days 90 (29th Oct)*	*Mean*
Control	1.85 ± 0.07^{Ab}	1.88 ± 0.04^{ACb}	$1.90 \pm .04^{ABCb}$	1.94 ± 0.05^{ABCb}	2.21 ± 0.04^{BCb}	2.31 ± 0.03^{Bb}	2.01
Chitin	1.81 ± 0.03^{Ab}	1.89 ± 0.05^{Ab}	2.0 ± 0.05^{Ab}	2.05 ± 0.04^{Abc}	2.31 ± 0.04^{ABbc}	2.40 ± 0.03^{Bb}	2.07
Chitosan	1.78 ± 0.04^{ABa}	1.75 ± 0.02^{ABa}	1.71 ± 0.02^{Ba}	1.80 ± 0.03^{ABDa}	1.91 ± 0.03^{BCa}	1.93 ± 0.04^{CDa}	1.81
Levamisole	1.80 ± 0.03^{Aa}	1.77 ± 0.04^{Aa}	1.74 ± 0.04^{Aa}	1.86 ± 0.03^{ABa}	1.96 ± 0.04^{BCa}	2.0 ± 0.03^{CDac}	1.85

Values with same superscript (capital in a row and small in a column) do not differ significantly ($P>0.05$).

Relative Percentage Survival (RPS)

The mortality percentage was found highest (80.5 per cent) in the control (infected) group and lowest (13.88 per cent) in chitosan supplemented group. The relative percentage survival was found highest (82.75) in chitosan supplemented group and lowest (48.24) in chitin supplemented group.

Table 4.3: Specific Growth Rate (SGR) (per cent) (Mean ± SE)

Days/ Treatments	Day 15 (15th Aug)	Day 30 (30th Aug)	Day 45 (14th Sep)	Day 60 (29th Sep)	Day 75 (14th Oct)	Days 90 (29th Oct)	Mean
Control	2.5 ± 0.11^{Ea}	2.37 ± 0.06^{DEa}	2.12 ± 0.08^{BDa}	1.90 ± 0.05^{BCa}	1.69 ± 0.05^{ACa}	1.49 ± 0.06^{Aa}	2.01
Chitin	2.60 ± 0.07^{Ea}	2.10 ± 0.09^{Db}	1.90 ± 0.07^{BDa}	1.80 ± 0.06^{BCa}	1.60 ± 0.08^{ACa}	1.40 ± 0.07^{Aa}	1.90
Chitosan	3.38 ± 0.06^{Eb}	3.0 ± 0.07^{Dc}	2.90 ± 0.09^{Cb}	2.50 ± 0.08^{Bb}	2.30 ± 0.09^{Bb}	1.90 ± 0.08^{Ab}	2.67
Levamisole	3.03 ± 0.10^{Ec}	2.6 ± 0.08^{Dd}	2.40 ± 0.07^{CDc}	2.31 ± 0.06^{BCb}	2.16 ± 0.10^{Bb}	1.78 ± 0.07^{Ab}	2.38

Values with same superscript (capital in a row and small in a column) do not differ significantly ($P>0.05$).

In the present study, when the fish were intra-peritoneally challenged with *Aeromonas hydrophila* on 30th and 58th day, the RPS of the fish supplemented with chitosan and levamisole was significantly higher ($p<0.05$) than the chitin supplemented group. This might be due to the enhancement of the non-specific immune system of the fish by chitosan and levamisole. Similar findings were reported by Gopalakannan and Venkatesan (2006) while challenging *Cyprinus carpio* with *Aeromonas hydrophila.* Baba *et al.* (1993) also reported that survival rate after challenging the fish with *Aeromonas hydrophila* was enhanced in common carp treated with levamisole. The RPS was significantly increased in *Cyprinus carpio* injected with 500 μg and 1 mg of glucan per fish against intra-peritoneal challenge of *Aeromonas hydrophila.*

Levamisole has also been found to be a possible modulator of the immune response of *Cyprinus carpio* (Siwicki, 1987, 1989; Baba *et al.* (1993) and rainbow trout, *Oncohrynchus mykiss* (Kajita *et al.,* 1990). After treatment with levamisole, both the fish species showed enhanced non-specific immune response activities and resistance to experimental challenge with pathogenic bacteria. Mortality percentage in the levamisole supplemented group was only 31 per cent and 88 per cent in the control (infected) group when all the groups were challenged with *Vibrio anguillarum.* However, in the present study lower RPS in the chitin supplemented group could be attributed to suppression of non-specific immune system of fish, as they might have developed tolerance to chitin administration. Lower RPS in chitin (1 per cent) supplemented group is also reported by Gopalakannan and Venkatesan (2006).

Table 4.4: Relative Percentage Survival (RPS) (per cent)

Treatment	Total No. of Fishes	No. of Mortalities	Survival Percentage	Mortality Percentage	RPS
Control	60	48	20.00	80.00	–
Chitin	60	25	58.33	41.66	47.90
Chitosan	60	5	91.67	8.33	89.60
Levamisole	60	9	85.00	15.00	81.30

Fish Production

The average daily gain in weight (g/day) was found higher in the chitosan supplemented group (0.79) followed by levamisole supplemented group (0.66), control (infected) group (0.42) and chitin supplemented group (0.40). At the stocking density of 16,666 fingerlings / ha which is almost double of what is recommended for semi-intensive carp culture, the maximum net fish production (kg /ha/8 months) was recorded in chitosan supplemented group (2896.3) showing the better FCR (1.81) and

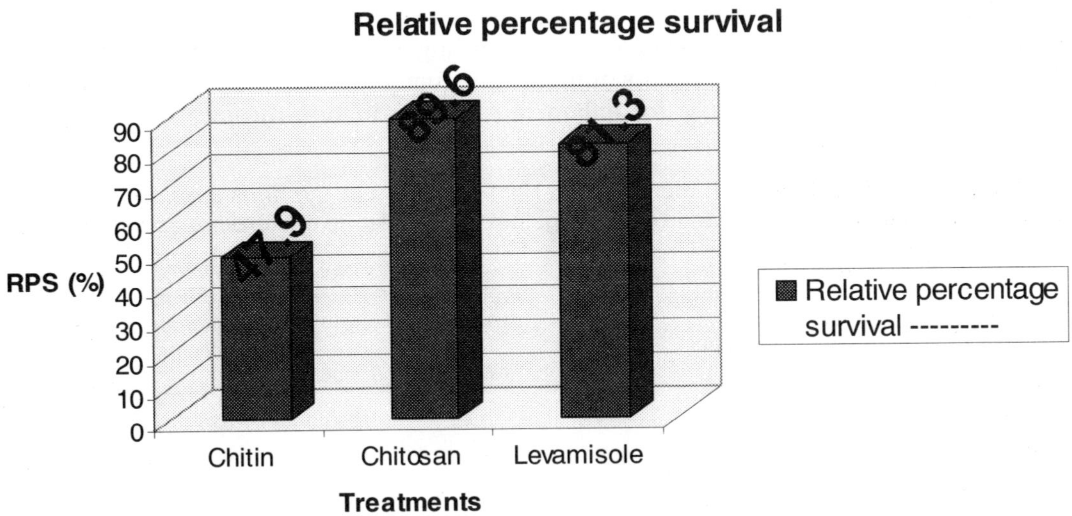

Figure 4.1: Relative Percentage Survival of Different Treatment Groups Challenged with *Aeromonas hydrophila*

survival (91.6 per cent). In the control (infected) group, the huge loss due to the mortality caused by *Aeromonas hydrophila* infection resulted in less production (336.0 kg/ha/8months) and less growth rate (0.42).

Results obtained in the present investigation revealed that the highest net fish production of 2896.3 kg/ha for 8 months of rearing period was achieved in chitosan supplemented group while the lowest of 336 kg/ha/ 8 months was recorded in control (infected) group. The production to this tune under monoculture of common carp is first time achieved in temperate climatic conditions of Kashmir valley where the growth of fish is hampered mainly during the winter period which extend for at least 3-4 months.

Gopalakannan and Venkatesan (2006) after conducting the monoculture of common carp for a period of 3 months demonstrated that the chitosan fed fish achieved maximum weight gain of 87.52g when compared to control group (40g) under tropical climatic conditions. However, in the present study, the common carp achieved maximum weight gain of 71.1g in chitosan supplemented group as compared to control (37.8g) under temperate climatic conditions in a rearing period of 3 months.

Based on the findings of the present investigation, it can be concluded that administration of the chitin (1 per cent), chitosan (1 per cent) and levamisole (250mg/kg of diet) in the fish diets certainly enhances growth and improves the survival and overall production of common carp under temperate climatic conditions of Kashmir valley. Chitosan and levamisole have shown positive effect on the overall growth and production of *Cyprinus carpio* by the way of improved SGR and FCR values. However, chitin was less effective in enhancing the growth and survival of *Cyprinus carpio* when compared to chitosan and levamisole. The low production in the control group was mainly due to mass mortality caused by pathogenic strain of *Aeromonas hydrophila* which is prevalent in the fish farms mainly during hatchery and rearing phases of common carp in Kashmir valley.

This approach is more relevant for use against diseases for which vaccines are not currently available. However, the exact schedule for dosage according to the age group is to be determined for effective commercial use of the immunostimulant in aquaculture. It is hoped that this base line information will be of potential use for enhancing fish production through proper fish health management practices.

Table 4.5: Variations in Growth, Survival and Net Production of *Cyprinus carpio* in Different Immunostimulant Supplemented Groups Against the Challenge of *Aeromonas hydrophila*

Treatment	*Stocking Density (no.'s/ha)*	*Average Growth Rate (g/day)*	*Weight Gain (g)*	*Net Production (kg)/Pond for 3 Months*	*Net Production (kg)/ha for 3 months*	*Net Production/ ha (kg) for 8 months*	*Survival (Per cent)*
Control	16,666	0.42	453.6	0.45	126.00	336.0	20.00
Chitin	16,666	0.40	1260	1.26	360.00	960.0	58.33
Chitosan	16,666	0.79	3910.5	3.91	1086.11	2896.3	91.67
Levamisole	16,666	0.66	3029.4	3.02	838.88	2237.03	85.00

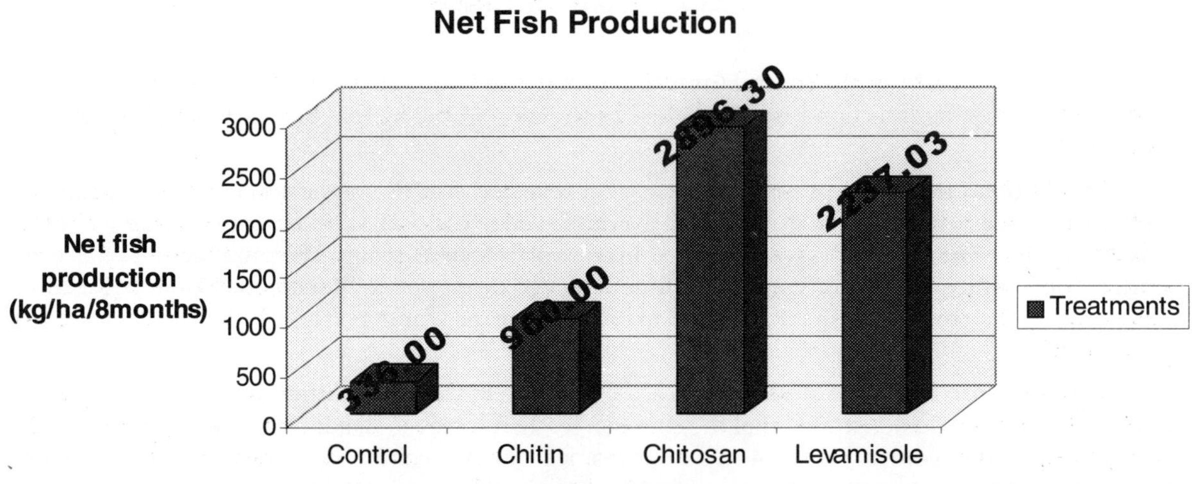

Figure 4.2: Net Fish Production of Different Treatments Against the Control

Acknowledgement

The author is thankful to Indian Council of Agricultural Research (ICAR) for providing financial assistance in the form of Junior Research Fellowship (JRF).

References

AOAC, 2006. *Official Methods of Analysis*, 18th edition, Horwitz W. Washington, DC, pp. 1018.

APHA, 1998. *Standard Methods for Examination of Water and Wastewater*, 20th edition. Lenore. S *et al.*, Washington, DC.

Alvarez Pellitero, P., Sitja Bobadilla, A., Bermudez, R. and Quiroga, M.I., 2006. Levamisole activates several innate immune factors in *Scophthalmus maximus*(L.) Teleostei. *Int. J. Immunopathol. Pharmacol.*, 19(4): 727–738.

Amend, D.F., 1981. Potency testing of fish vaccines. In: *Fish Biologies: Serodiagnostics and Vaccines Development in Biological Standardization*, (Eds.) D.P. Anderson and H. Hennessen. Karger, Basel, pp. 447–454.

Anderson, D.P., 1992. Immunostimulants, adjuvants, and vaccine carriers in fish: Applications to aquaculture. *Annu. Rev. Fish Dis.*, 2: 281–307.

Anderson, D.P. and Siwicki, A.K., 1994. Duration of protection against *Aeromonas salmonicida* in brook trout immunostimulated with glucan or chitosan by injection or immersion. *Progressive Fish Culturist*, 56(4): 258–261.

Baba, T., Watase, Y. and Yoshinaga, Y., 1993. Activation of mononuclear phagocytes function by levamisole immersion in carp. *Nippon Suisan Gakkaishi*, 59: 301–307.

Baulny, M.O.D., Quentel, C., Fournier, V., Lamour, F. and Gouvello, R.L., 1996. Effect of long-term oral administration of β-glucan as an immunostimulants or an adjuvant on some non-specific parameters of the immune response of turbot *Scophthalmus maximus. Disease of Aquatic organisms*, 26: 139–147.

Chabot, D.J. and Thunne, R.L., 1991. Protease of the *Aeromonas hydrophila* complex; identification, characterization and relation to virulence in channel cat fish, *Ictalurus punctatus. Journal of Fish Disease*, 14: 171–183.

Esteban, M.A., Cuesta, A., Ortuno, J. and Meseguer, J., 2000a. Effect of injecting chitin particles on the innate immune response of gilt head seabream (*Sparus auratus*). *Fish and Shellfish Immunology*, 12: 543–564.

Esteban, M.A., Cuesta, A., Ortuno, J. and Meseguer, J., 2000b. Immunomodulatory effects of dietary intake of chitin on gillhead seabream (*Sparus aurata L.*) innate immune system. *Fish and Shellfish Immunology*, 11 (2): 303–315.

Gopalakannan, A. and Venkatesan, A. 2006. Immunomodulatory effect of dietary intake of Chitin, Chitosan and Levamisole on the immune system of *Cyprinus carpio* and control of *Aeromonas hydrophila* infection in ponds. *Aquaculture*, 255: 179–187.

Heo Gang Joon, Kim Jeong Ho, Jeon Byong Gu, Park Ki Yeol, Ra Jeong Chan, Heo, G.J., Kim, J.H., Jeon B.G., Park, K.Y. and Ra, J.C., 2004. Effects of FST-Chitosan mixture on cultured rockfish (*Sebastes schlegeli*) and olive flounder (*Paralichthys olivaceus*). *Korean Journal of Veterinary Public Health*, 25(3): 141–149.

Hussain, S.A., Samoon, M.H., Najar, A.M., Balkhi, M.H. and Rashid, R., 2005. Occurrence of Fin Rot Disease in Common Carp (*Cyprinus carpio)* in Kashmir. *Journal of Veterinary Public Health*, 3: 79–81.

Janssen, P.A., 1976. The levamisole story. *Progress in Drug Research*, 20: 347–383.

Kim, K.W., Wang, X.J., Choi, S.M., Park, G.J., Koo, W.J. and Bai, S.C., 2003. No synergistics effects by the dietary supplementation of the ascorbic acid, α-tocopheryl acetate and selenium on the growth performance and challenge test of *Edwardsiella tarda* in fingerling Nile tilapia, *Oreochromis niloticus* L.*Aquaculture Research*, 34: 1053–1058.

Kono, M., Matsui, T. and Shimizu, C., 1987. Effect of chitin, chitosan and cellulose as diet supplement on the growth of cultured fish. *Nippon Suisan Gakkaishi*, 53: 125–129.

Kajita, Y., Sakia, M., Atsuta, S. and Kobayashi, M., 1990. The immunomodulatory effectof levamisole on rainbow trout *Onchorynchus mykiss*. *Fish Pathology*, 25: 93–98.

Misra, C.K., Das, B.K., Mukherjee, S.C. and Phalguni, P., 2005. Effect of long term administration of dietary β-glucan on immunity, growth and survival of *Labeo rohita* fingerling. *Aquaculture*, 255(1–4): 82–94.

Mulero, V., Esteban, M.A., Munoz, J. and Meseguer, J., 1998. Dietary intake of levamisole enhances the immune response and disease resistance of the marine teleost gilthead seabream (*Sparus aurata* L.). *Fish and Shellfish Immunology*, 8: 49–62.

Niki, L., Albright, L.J. and Evelyn, T.P.T., 1991. Influence of seven immunostimulants on the immune response of Coho salmon to *Aeromonas salmonicida*. *Diseases of Aquatic Organisms*, 12(1): 7–12.

Ricker, W.E., 1979. Growth rates and models. In: *Fish Physiology*, (Eds.) W.S. Hoar, P.J.Randall and J.R. Brett. Academic Press, New York, pp. 677–743.

Sakai, M., 1999. Current research status of fish immunostimulants. *Aquaculture*, 172(1–2): 63–92.

Secombes, C.J., 1994. Enhancement of fish phagocyte activity. *Fish and Shellfish Immunology*, 4: 421–436.

Shibata, Y., Metzger, W.J. and Myrvik, Q., 1997. Chitin particle-induced cell mediated immunity is inhibited by soluble mannan. *Journal of Immunology*, 159: 2462–2467.

Shiau, S.Y. and Yu, Y.P., 1999.Dietary supplementation of chitin and chitosan depressed growth in Tilapia, *Oreochromis niloticus x O. auratus*. *Aquaculture*, 179: 439–446.

Siwicki, A.K. and Korwin-Kossakomski, M., 1988. The influence of levamisole on the growth of carp (*Cyprinus carpio*) larvae. *Journal of Applied Ichthyology*, 4(4): 178–181.

Siwicki, A.K., 1987. Immunomodulating activity of levamisole in carp spawner. *Journal of Fish Biology*; 31 (Supplement A): 245–246.

Siwicki, A.K., 1989. Immunomodulating influence of levamisole on non–specific immunity in carp (*Cyprinus carpio*). *Development and Comparative Immunology*, 13 : 87–91.

Stickney, R. Robert, 2000. *Encyclopedia of Aquaculture*, p. 676–679.

Siwicki, A.K., Anderson, D.P. and Rumsey, G.L., 1994. Dietary intake of immunostimulants by rainbow trout affects non-specific immunity and protection against furunculosis. *Veterinary Immunology and Immunopathology*, 41(1–2): 125–139.

Yadav, M., Indira, G. and Ansary, A., 1992. Cytotoxin elaboration by *Aeromonas hydrophila* isolated from fish with epizootic ulcerative syndrome. *Journal of Fish Disease*, 159: 183–189.

Chapter 5

Estimation of Activity of Some Metabolic Enzymes in the Fish *Cyprinus carpio* as a Function of their Age in Juveniles

☆ *Meenakshi Jindal, K.L. Jain, R.K. Verma and Simmi*

Introduction

Global fisheries is facing constant decline in fish population, both in coastal and inland water resources on account of continuously increasing water pollution. Growth in fishes and pollution are the inter-related terms. As the growth and size of fish increases the impact of pollutants also increases, however, the sensitive stage of fish such as the juvenile stage are highly concentrated to such concerned impact of pollutants. The liver is an important organ involved in metabolic processes as well as in detoxifying the toxic chemicals called as xenobiotics. Liver enzymes more importantly the metabolic enzymes are the health determinants commonly applied in studding toxicity and fish health. There is a need to correlate their activity with early growth in fish which could assure their correct application in clinical tests. The early growth stages of fish are generally employed in testing animals for various toxicological studies. Pelletier *et al*. (1993) showed a strong positive correlation between growth rate and the activity of some glycolytic enzymes like lactate dehydrogenase (LDH), pyruvate kinase (PK) and phosphofructokinase (PFK) in white muscle of Atlantic cod (*Gadus morhua*).

Materials and Methods

Healthy fish samples of approximately 3 to 5cm of freshwater teleost *C.carpio* (common carp) were procured in the months of June to October from the culture maintained in a controlled fish farm at Hindwaan village near Hisar. Their length and weight were measured to calculate the condition factor and the fish were anesthetized with MS-222 to dissect out their liver for the analysis of the activity of the enzymes Succinate dehydrogenase (SDH), Pyruvate deydrogenase (PDH) and Glutamate dehydrogenase (GDH).

The activity of these enzymes were estimated following the methods given by Nachlas *et al.*, 1960, by extracting a stock solution at a concentration of 5 mg per ml (wet weight) in 0.1 M phosphate buffer (pH 7.7). Reaction mixture was prepared by adding 0.5 ml of sodium succinate, sodium pyruvate and sodium glutamate for SDH, GDH and PDH, respectively. Mixture was treated sequently with phosphate buffer, gelatin and then the respective enzyme and PMS. Absorbance was recorded within 30-60 minutes at 540nm. The blank contained all the reagents except the substrate, which was replaced by sodium fumarate. Varying amounts of INT (0.04 to 0.24 mg) were used for preparing standard curve. DNPH was used as reducing agent. The selected dehydrogenase enzyme activity was expressed in terms of µ moles of formazan formed/ mg protein /h.

Results and Discussion

Study of age related changes in the activity of metabolic enzymes, Succinate dehydrogenase (SDH), Glutamate dehydrogenase (GDH) and Pyruvate dehydrogenase (PDH) in liver (Table 5.1) showed a significant variation with different growth periods. The enzyme activity of SDH in liver increased rapidly with age from 70 days old fish to 100 days old fish, as indicated by the mean values, *i.e.* 0.244 µg of formazan formed/mg of protein/h in 70 days and 0.394 µg in 100 days of growth which is almost double. Activity of GDH showed an increase from 0.199 µg in 70 days old fish to 0.229 µg of formazan /h in 100 days old fish. PDH also showed a significant increase in activity in the same way. The mean values were 0.205 µg of formazan formed/mg of protein/h in 40 days and 0.240 µg of formazan formed in 70 days, with an increase of 17.1 per cent.

Table 5.1: Age Related Gain in Fish Length and Mass and Changes in Activity of Succinate Dehydrogenase (SDH), Glutamate Dehydrogenase (GDH), Pyruvate Dehydrogenase (PDH) (µg of formazan formed/mg of protein/h) in liver of *Cyprinus carpio*

Age (Days)	*Per cent Length Gain/Day*	*Per cent Mass Gain/Day*	*Final Condition Factor*	*SDH*	*GDH*	*PDH*
40	7.0	6.4	9.2	0.196	0.184	0.205
70	8.5	9.3	6.1	0.244 (24.5)	0.199 (8.15)	0.240 (17.1)
100	10.3	15.1	6.8	0.394 (101.0)	0.229 (24.4)	0.268 (30.7)
130	10.85	16.6	5.0	0.503 (156.6)	0.258 (40.2)	0.321 (56.5)
CD (5 per cent)				0.045	0.008	0.014
F-cal				100.92	200.07	126.0

* Level of significance is $p<0.0001$.

These showed a strong relationship of their activity with growth rate both in length and mass, higher than with the condition factor (Table 5.2), not withstanding the condition factor also increased significantly with age. SDH in liver, varied with growth in length by 89 per cent and 87 per cent due to growth in mass. GDH also contributed to the same per cent of variability in growth rate in length as well as in mass.

Gain in length involve different physiological processes than gain in mass, as the former employees synthesis of structural molecules, while the important part of gain in mass may be due to the restoration and accumulation of energy resources. Changes in enzymatic activities as evident in this study could be well correlated with metabolic requirement of the fish as per the growth with age. The study evidenced highest increase in SDH, GDH and PDH enzyme activities in 130 days old fish, showing an ultimate increase up to 156.6, 40.0 and 56.5 per cent, respectively.

Table 5.2: Linear Regression Between the Dehydrogenases Enzyme Activity (units/g wet tissue) and Growth in Length, Mass and Condition Factor in Liver of Fish *Cyprinus carpio*

Variables	*Growth in Length (Per cent gain/day)*	*Growth in Mass (Per cent gain/day)*	*Condition Factor*
Succinate dehydrogenase	Y=-0.390+0.079GL (R^2=0.889)*	Y=-0.007+0.029GM (R^2=0.873)	Y=0.725-0.232CF (R^2=0.521)
Glutamate dehydrogenase	Y=0.047+0.0185GL (R^2=0.894)	Y=0.047+0.018GM (R^2=0.894)	Y=0.312-0.056CF (R^2=0.554)
Pyruvate dehydrogenase	Y=0.011+0.127GL (R^2=0.863)	Y=0.143+0.009GM (R^2=0.833)	Y=0.484-0.089CF (R^2=0.641)

*: Level of significance is $p<0.0001$.

Age and growth related changes in liver composition and metabolic enzymes activities are found to be interlinked with toxicity in various water resources. Pelletier *et al.* (1993) have shown a strong positive correlation between growth rate and the activity of the glycolytic enzymes lactate dehydrogenase (LDH), pyruvate kinase (PK) and phosphofructokinase (PFK) in white muscle of Atlantic cod (*Gadus morhua*). Activities in muscle of the glycolytic enzymes lactate dehydrogenase (LDH) and pyruvate kinase (PK) increased significantly with increasing body size.

CF (mass/volume relationship) is often used to monitor changes in population health in wild fish. The physiological condition of fish in these natural habitats is of increasing concern because of continuous decline in the abundance of fish species. Simmi (2007) also evidenced a relationship in between HSI, liver composition and the period of growth in *C. carpio* juveniles. Determining changes in body index parameters and liver, red muscle and white muscle enzyme profiles in fed and four month starved fish plaice, *Pleuronectes platessa* (Moon and Johnston, 1980) showed the liver possessing lowest glycolytic activity, but highest gluconeogenic capacity of the three tissues. The increase and decrease in the activity of metabolic enzymes is a clear evidence of changes in the metabolic environment of the fish may be due to age or stress as both these parameters influence dietary intake. Gary *et al.* (2001) studied the activities of oxidative and glycolytic enzymes show body size-dependent relationships across a wide variety of taxa. Sastry and Sharma (1980) also observed increase in the level of serum enzymes namely glutamate-oxaloacetate transaminase and glutamate-pyruvate transaminase after acute stress on account of exposure to mercury in *Channa punctatus*. Increased activities of these enzymes may be due to the damage of liver.

Conclusion

The increase and decrease in the activity of metabolic enzymes in fish liver as indicated in this study is a clear evidence of changes in the metabolic environment of the fish as affected by its age since it influence dietary intake. as well as change in body structure and function. In conclusion this study provide a good evidence to apply these enzymes as an index for estimating fish age in their juvenile stage and to correlate correctly with any health parameter in toxicological studies.

References

Gary, P., Leary, B., Peter, H.C., Hochachka, W. and Moyes, C.D., 2001. *Histochemistry and Cytochemistry*, 49: 1025–1032.

Moon, T.W. and Johnston, I.A., 1980. Starvation and the activities of glycolytic and gluconeogenic enzymes in skeletal muscles and liver of the plaice, *Pleuronectes platessa. J. Comp. Physiol.*, 136: 31–38.

Nachlas, M.N., Margulies, S.I. and Seligman, A.M., 1960. A clorimetric method for the estimation of succinate dehydrogenase activity. *J. Biol. Chem.*, 235: 499–503.

Pelletier, D., Gurderely, H., and Dutil, J.-D., 1993. Effect of growth rate, temperature and season on glycolytic enzyme activities in white muscle of cod *Gadus morhua. J. Exp. Zool.*, 265: 477–487.

Sastry, K.V. and Sharma, S.K., 1978. The effect of *in-vivo* exposure of endrin on the activities of acid alkaline and Glucose-6-phospatase in liver and kidney of *Ophiocephalus* and *C. punctatus. Bull. Enviorn. Contain. Toxicol.*, 20(4): 456–460.

Simmi, 2007. Age dependent changes ion liver composition and enzyme activity in *C. carpio. M.Sc. Thesis,* CCS HAU, Hisar.

Chapter 6

Studies on Oxygen Levels and Temperature Fluctuation in Dhanegaon Reservoir of Maharashtra

☆ *M.V. Lokhande, D.S. Rathod, V.S. Shembekar and K.G. Dande*

Introduction

Lakes, reservoirs, rivers and streams have been critical to the establishment of civilizations throughout human history. Water bodies are essential to humans not only for drinking purposes but also for transportation, irrigation, energy production, industry and west disposal. About 80 per cent of earth surface is covered by water but the inland fresh water availability is account for less than one per cent.

Today many reservoirs, rivers and lakes in India contaminated run off water from expanding urban and agricultural areas, air pollutants, and hydrologic modification such as drainage of wet lands are just few of the many factors that continue to degrade surface waters. Water is needed for our life at every activities it has become the first responsibility to maintain the quality of water and to conserve the fresh water aquatic environment.

The present investigation has been undertaken to study the various parameters of Dhanegon reservoir which is within the permissible limit prescribed by WHO and ICMR.

The Dhanegaon reservoir is large sized reservoir constructed across Manjara river near village Dhanegaon in Osmanabad district of Maharashtra. It is basically constructed for irrigation purpose.This reservoir water supply to Latur, Kaij, Ambajogai and Kallamb for drinking purposes. The reservoir constructed in 1980 lies between 1823 to 1855 N latitude, 7515 to 7615 E longitude. Dhanegaon reservoir has a catchment area of about 2371 Sq. Km. The 73 villages of Beed, Osmanabad and Latur is suppose to be beneficiaries of this project.

Materials and Methods

Water samples are collected in separate wide mouthed crew capped, air tight, opaque polythene container on monthly basis for a period of one year June-2003 to May 2004 at three sampling stations named as spot A, B and C. The spot A is located one side of reservoir and spot B located at the Dhanegaon camp near about 6 Km. away from spot A the spot C is located in the reservoir.

The temperature of lake water samples was recorded in the field itself with the help of centigrade thermometer. The amount of dissolved oxygen in the collected water samples was estimated by Winkler's titrometric method in the laboratory (APHA, 1995 and IAAB, 1998).

Results and Discussion

In the present investigation the maximum temperature of Dhanegaon reservoir was recorded in May it was 30.0ºC, 30.4ºC and 30.3ºC at spot A, B and C respectively in the year 2003-2004.The fluctuation pattern of water temperature is expressed in Table 6.1 and graphically represented in Figure 6.1. The temperature fluctuate in month to month but the seasonal fluctuation of temperature the minimum temperature was recorded in the rainy season and maximum temperature was recorded in the summer season but moderate recorded in the winter season. The temperatures of any area vary from sunlight to shade and from daylight to dark. In Dhanegaon reservoir the maximum and minimum temperature recorded is depend upon the brightness and shadow ness of sunshine duration. The water temperature is determining factor for the distribution of aquatic organisms (Allern, 1920) and the variation in water temperature may due to different timing of collection and the influence of season Jayaraman *et. al.* (2003). In Badkhal lake Haryana the maximum temperature recorded was 31.2 °C and minimum was 15.0ºC (Kaushal and Sharma 2001). Sedemkar and Angadi (2003) recorded temperature in the range of 20.30 to 30.90ºC in Jagat tank and Pala tank. A positive correlation was observed between water temperature and dissolved oxygen. Shastri (2000) recorded water temperature ranges between 18 to 29 °C. Surve *et.al.* (2005) recorded the water temperature of Barul dam. It varied between 22.2 to 33.0 °C, 22 to 32.9ºC and 22.3 to 33.0ºC at site I, II and III respectively.

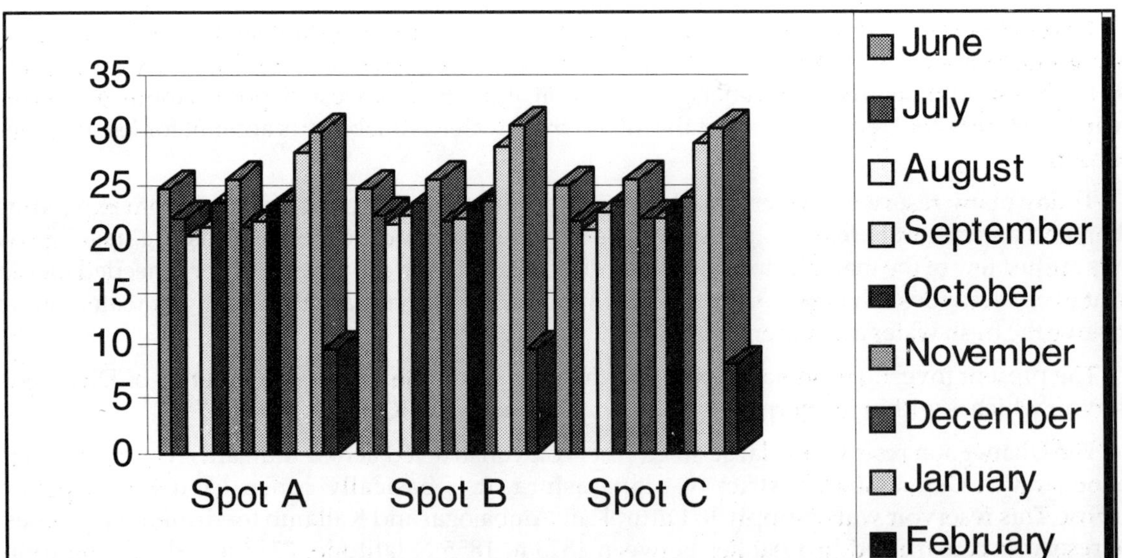

Figure 6.1: Monthly Variation in Water Temperature

Table 6.1: Monthly Mean Values of Temperature from Dhanegaon Reservoir Water Samples during the Year 2003–2004

Months	*Spot A*	*Spot B*	*Spot C*
June	24.6	24.8	25.0
July	21.8	22.2	21.7
August	20.3	21.4	20.8
September	21.1	22.3	22.4
October	23.2	23.4	23.5
November	25.4	25.4	25.5
December	21.2	21.6	21.8
January	21.7	21.8	21.9
February	22.9	23.0	22.7
March	23.5	23.7	23.8
April	28.1	28.5	28.9
May	30.0	30.4	30.3

All values are expressed in °C.

In the present investigation the amount of dissolved oxygen was found to be maximum in January in the year 2003-2004 with the values 12.6 mg/l, 12.4 mg/l and 13.5 mg/l at spot A, B and C respectively. The amount of dissolved oxygen level were minimum in June at spot A 8.2 mg/l, 7.2 mg/l at spot B in June but 7.6 mg/l at spot C in the month of May. Monthly mean values of dissolved oxygen are expressed in Table 6.2 and graphically represented in Figure 6.2. The seasonal fluctuation of dissolved oxygen is observed maximum in winter season and minimum in summer season. This is indicates the positive correlation of dissolved oxygen and water temperature. Increase the temperature of water in summer is as result the minimum amount of dissolved oxygen. Surve *et.al.* (2005) reported that the dissolved oxygen fluctuated between 3.2 and 7.6 mg/l, 3.4 to 7.9 mg/l and 3.2 to 7.6 mg/l at S1, S2 and S3 sites respectively.

The higher values of dissolved oxygen recorded in the winter season due to high solubility at low temperature and less degradation of organic substances in winter. The low oxygen in water can kill the fish and other organisms present in water. The potable range of drinking water of dissolved oxygen is above the 4.0 mg/l. In the present investigation was observed the all value of dissolved oxygen within the permissible limit prescribed by WHO and ICMR.

The dissolved oxygen is one of the most important parameter in water quality and is an index of physical and biological processes going on in water. There are two main sources of dissolved oxygen in water, diffusion from air and photosynthetic activities within the water. Diffusion of oxygen from atmosphere is physical phenomenon and depends upon solubility of oxygen, which often effected by factors like temperature, water movements and salinity. Pulle *et.al.* (2005) recorded dissolved oxygen was ranging from 3.08 to 4.28 mg/l, 3.30 to 6.24 mg/l and 4.78 to 6.28 mg/l at three sites I, II and III respectively. The lowest value was observed during summer while highest value during winter months is similar finding in Dhanegaon reservoir. Sedamkar and Angadi (2003) recorded dissolved oxygen minimum 5.2 mg/l in the month of January and maximum 10.4 mg/l in the Pala tank near Gulbarga. A positive correlation was observed between water temperature and dissolved oxygen stated by Jha

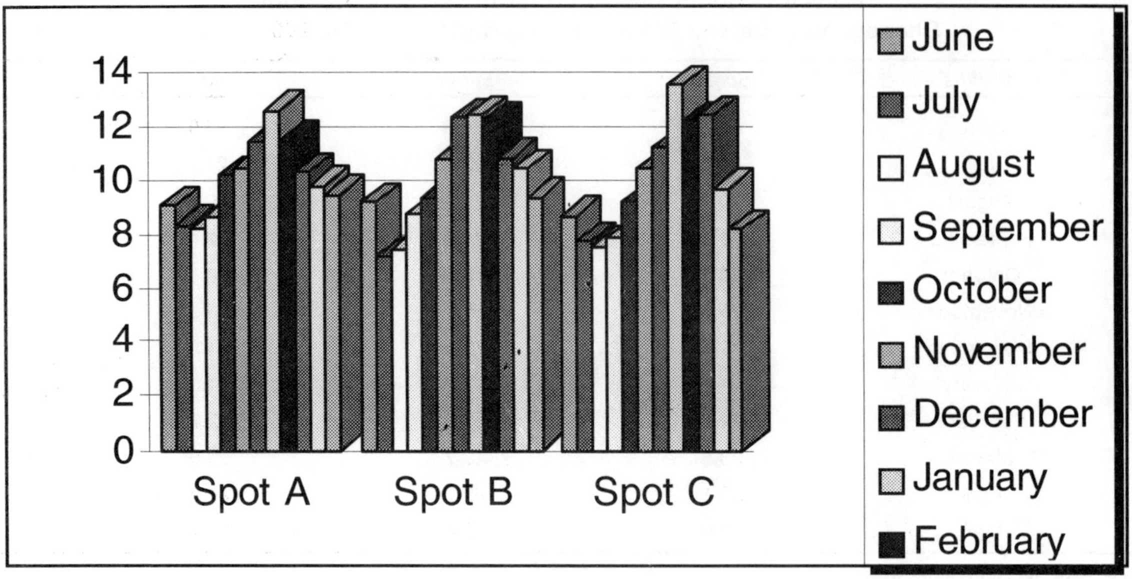

Figure 6.2: Monthly Variation of Dissolved Oxygen from Dhanegaon Reservoir Water Samples During the Year 2003–2004

et.al. (2003). The dissolved oxygen varied between 5.2 mg/l and 12.6 mg/l. It was increased from early may onwards and reached the peak at the end of August.

Table 6.2: Monthly Mean Values of Dissolved Oxygen from Dhanegaon Reservoir Water Samples During the Year 2003–2004

Months	*Spot A*	*Spot B*	*Spot C*
June	9.1	9.2	8.7
July	8.4	7.2	7.8
August	8.2	7.5	7.6
September	8.7	8.8	7.9
October	10.2	9.4	9.2
November	10.5	10.8	10.4
December	11.5	12.3	11.2
January	12.6	12.4	13.5
February	11.5	12.2	12.2
March	10.3	10.8	12.4
April	9.8	10.4	9.7
May	9.5	9.4	8.2

All values are expressed in mg/l.

Raghuwanshi *et.al.* (2005) recorded dissolved oxygen was ranging from 12.8 mg/l to 18.6 mg/l at lower lake Bhopal. The minimum value recorded in the month of April while maximum value recorded in the month of June. Radhika *et.al.* (2004) studied on the Vallayani lake and reported that the dissolved oxygen in surface water ranging from 4.83 to 7.11 mg/l, 4.11 to 6.81 mg/l and 4.71 to 6.80 mg/l during pre-monsoon, monsoon and post-monsoon respectively. No significant variations in dissolved oxygen were encountered both spatially and temporally in the water of Vallayani Lake.

In the present investigation the positive correlation between the water temperature and dissolved oxygen. The temperature is increases the amount of dissolved oxygen is decreases and the temperature is decreased the amount of dissolved oxygen increased in Dhanegaon reservoir.

Acknowledgements

The authors are thankful to Principal Dr. R. L. Kawale and Dr. D. G. Salunke, Head, Department of Zoology and Fishery Science, Rajarshi Shahu College, Latur for providing necessary facilities.

References

APHA, 1995. *Standard Methods for Examination of Water and Wastewater 19th Edn.* American Public Health Organization, American Water Works Association/Water Pollution Control Federation, Washington, D.C.

ICMR, 1995. *Manual of Standard of Quality of Drinking Water Supplies,* special report series No. 44, 2nd Ed.

Jayaraman, P.R., Gangadevi, T. and Nayer Vasudevan, T., 2000. Water quality studies on Karamana River Thiruvananthapuram District, South Kerala, India. *Poll. Res.*, 22(1): 89–100.

Jha, Prithviraj and Sudip, Barel, 2003. Hydrobiological study of lake Mirik in Darjelling, Himalaya. *J. of Env. Biol.*, 4(3): 339–344.

Kodarkar, M.S., Diwan, A.D., Murugan, N. Kulkarni, K.M. and Anuradha Ramesh, 1998. Methodology for water analysis: Physico-chemical, biological and microbiological. Indian Association of Aquatic Biologists, Hyderabad.

Pulle, J.S., Khan, A.M., Ambore, N.E., Kadam, D.D. and Pawar, S.K., 2005. Assessment of ground water quality of Nanded city. *Poll. Res.*, 24(3): 657–660.

Radhika, C.G., Mini, I. and Gangadevi, T., 2004. Studies on abiotic parameters of tropical fresh water lake–Vellayani lake, Thiruvananthapuram, Dist. Kerala. *Poll. Res.*, 23(1): 49–63.

Raghuwanshi Arun, K., 2005. The impact of physico-chemical parameter of lower lake Bophal on the productivity of Eichhornia Carassipes. *Eco. Env and Cons.*, 11(3–4): 333–336.

Sedamkar Eswarlal and Angadi, S.B., 2003. Physico-chemical parameters of two fresh water bodied of Gulbarga, India with special reference to phytoplankton. *Poll. Res.*, 22(3): 411–422.

Surve, P.R., Ambore, N.E. and Pulle, J.S., 2005. Correlation co-efficients of some physico-chemical characteristics of Barul dam water, District. Nanded (M.S.). *Poll. Res.*, 24(3): 653–656.

Chapter 7

Seasonal Variation in Turbidity, Total Solids, Total Dissolved Solids and Total Suspended Solids of Dhanegaon Reservoir in Maharashtra

☆ *M.V. Lokhande, D.S. Rathod, V.S. Shembekar and S.V. Karadkhele*

Introduction

Life in aquatic environment is largely governed by physico-chemical characteristics and their solubility all process in water bodies are governed by radiation directly or indirectly. The turbidity is defined as influenced to the passage of light by suspended particles in water it measured the effect on the transmission of light. Turbidity makes the water unfit for domestic purpose and if it is caused by suspended particles it absorb considerable amount of nutrients like phosphate, potassium and nitrogen making them unavailable for phytoplankton. The turbidity may be caused by a wide variety of suspended materials ranged from colloidal to coarse dispersion makes water unfit for drinking according to Katariya (1994). The quantum of radiation absorbed by water is greatly influenced by physico-chemical properties such as intensity of light at the surface, difference in latitude, suspended particles present in water, dissolved organic compound in water (Wetzel and Likes, 1979, Hutchinson, 1975). The high level of turbidity may minimize the phytoplankton population was observed by Eddy (1934). The total solids are sum of values of total dissolved solids and total suspended solids. The direct relationship between rainfall and total solids was attributed to an increased load of soluble salt from the catchments areas as a result of surface run-off. Total solids are resides left after evaporation of the unfiltered water sample at a specific temperature. These residues include both suspended and dissolved solids. Swarnalatha and Narsing Rao (1997) showed the interesting relationship between temperature and rainfall. The lower concentration was observed during winter. Generally high total solids are present in water body during summer season turn to the effect of temperature. The pollution has direct relationship with dissolved solids Verma *et.al.* (1997). Walling (1980) and Goltman (1975) showed the inverse relationship between dissolved solids and discharge of water in the river. The total suspended solids ate the cause of suspended particles inside the water body influencing turbidity and transparency Prasad and Sexena (1980).Regina and Nabi (2004) reported the disposal sewage and industrial pollutant, contribute suspended matter to rivers and streams. Sharma *et. al.* (1981) observed the total suspended solids greater than the turbidity, which may be due to the gradual sedimentation on the bottom.

The present investigation was carried out on the study of seasonal variation of turbidity, total solids, total dissolved solids, and total suspended solilds of Dhanegaon reservoir. The Dhanegaon reservoir is large sized reservoir constructed across Manjara river near village Dhanegaon in Osmanabad district of Maharashtra. It is basically constructed for irrigation purposes. About 73 villages of Beed, Osmanabad and Latur are supposed to beneficiaries of this project.

Materials and Methods

Turbidity of water was measured by Secchi disc having diameter 20 cms and divided in to black and white quadrants. The depth at which secchi disc was disappear was determined. The turbidity of water was measured at the time of sampling early in the morning hours. The total solids, Total dissolved solids, and total suspended solids were determined by standard methods suggested by APHA (1985) and IAAB (1998). Water samples for the analysis were collected from three stations namely Spot A, B and C from the reservoir of every months during the year 2003-2004.

Results and Discussion

The monthly variation of TS, turbidity TDS and TSS are given in Table 7.1 and graphically represented in Figures 7.1, 7.2, 7.3 and 7.4.

Table 7.1: Seasonal Variation in Total Solids, Turbidity, Total Dissolved Solids and Total Suspended Solids of Dhanegoan Reservoir Water (2003–2004)

Months	*Total Solids*			*Turbidity*			*Total Dissolved Solids*			*Total Suspended Solids*		
Spots	*A*	*B*	*C*	*A*	*B*	*C*	*A*	*B*	*C*	*A*	*B*	*C*
June	370	410	490	78	80	82	235	275	360	135	135	130
July	460	430	560	90	107	110	320	270	405	140	160	155
August	530	530	590	110	130	118	365	350	420	165	180	170
September	470	550	560	100	105	98	340	415	390	130	135	170
October	326	300	370	92	83	75	195	170	245	131	130	125
November	240	310	310	98	97	98	145	190	190	95	120	120
December	200	210	250	94	112	105	100	105	145	100	110	105
January	330	340	360	87	86	85	220	230	250	110	110	125
February	340	350	390	85	80	80	230	230	265	110	120	120
March	380	410	450	88	84	74	255	290	330	125	420	120
April	370	430	440	75	84	74	250	300	315	120	130	125
May	430	390	450	80	82	85	300	270	330	130	120	120

All values are expressed in mg/l. except turbidity in cms.

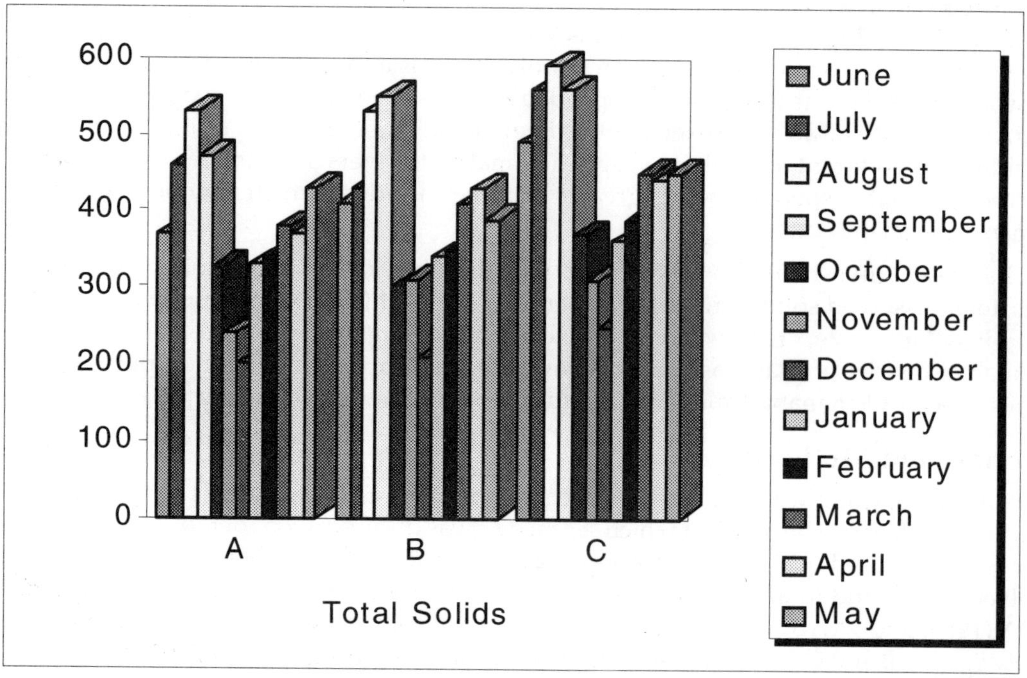

Figure 7.1: Monthly Variation in Total Solids

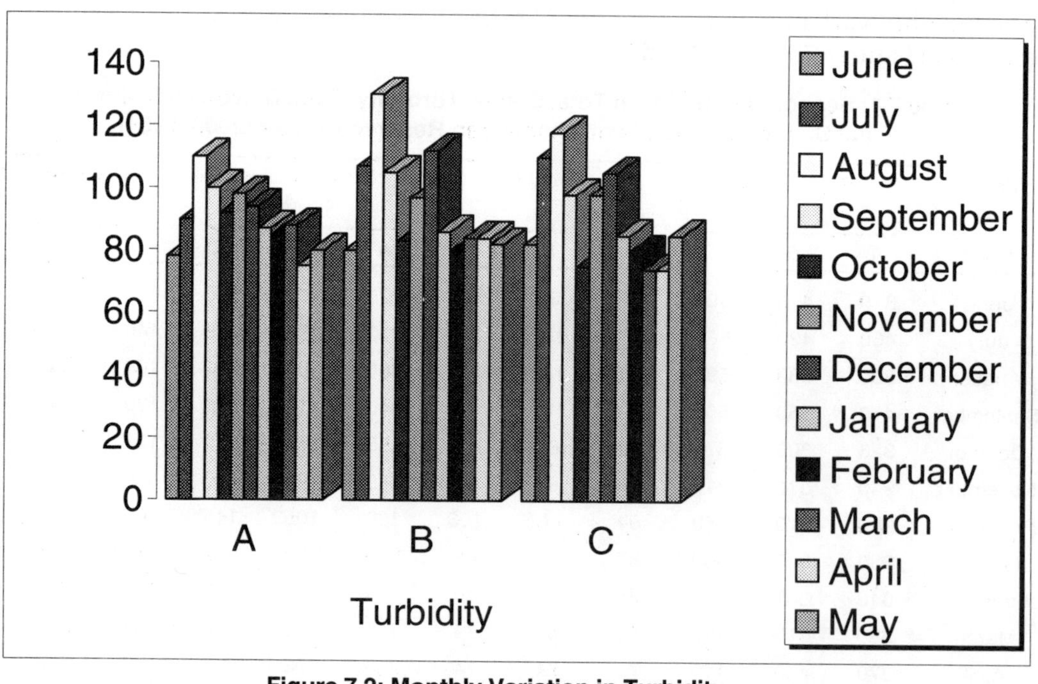

Figure 7.2: Monthly Variation in Turbidity

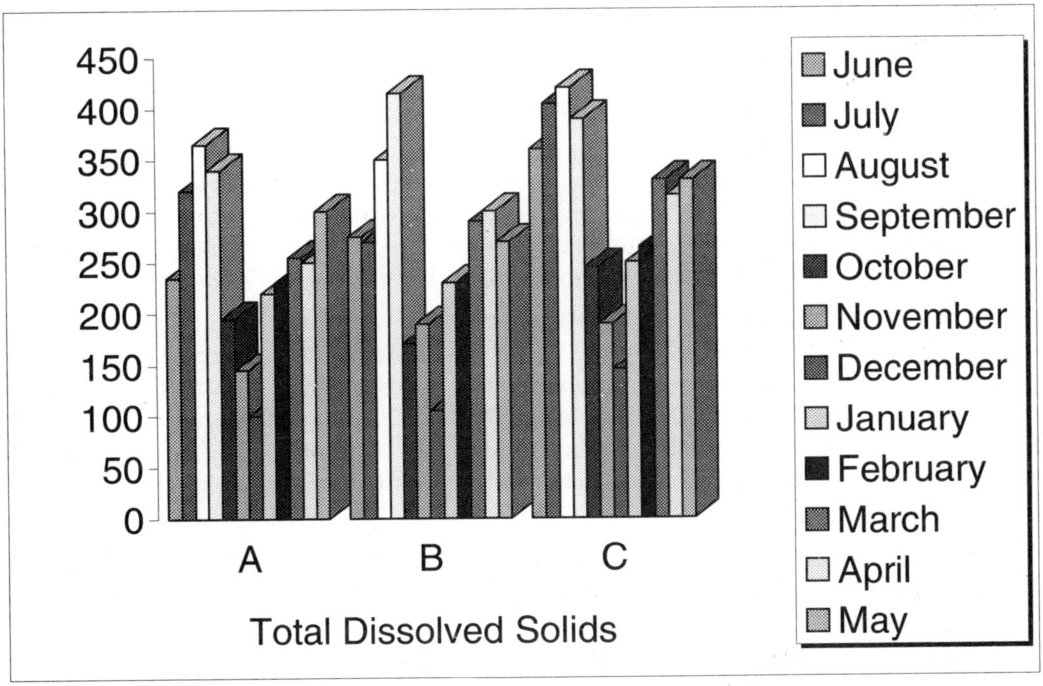

Figure 7.3: Monthly Variation in Total Dissolved Solids

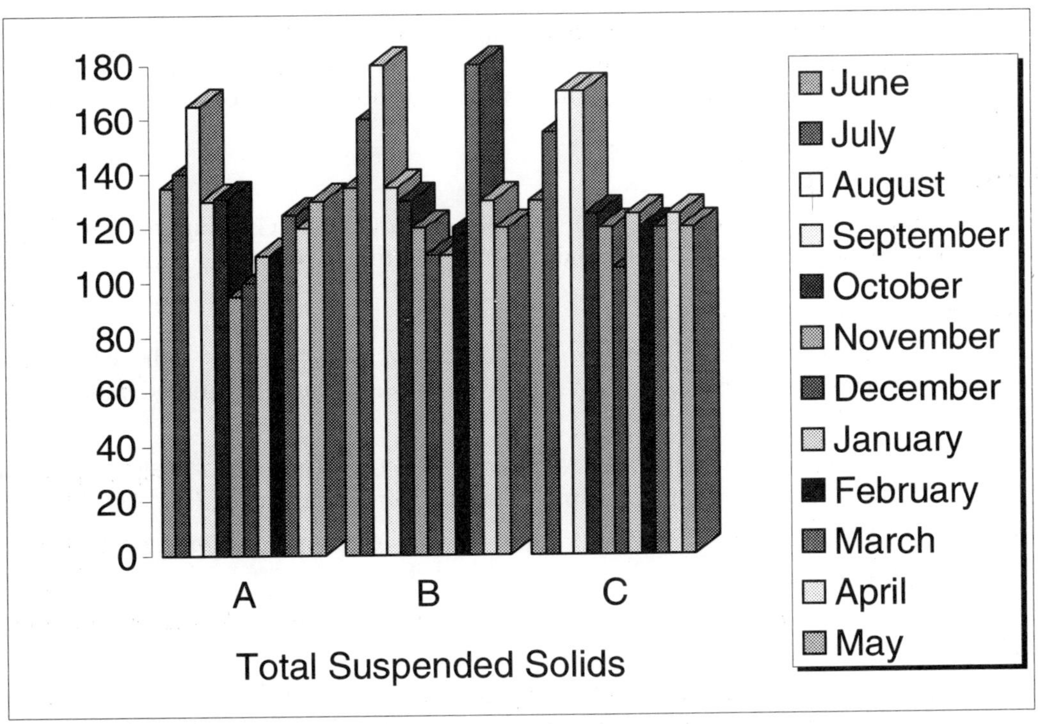

Figure 7.4: Monthly variation in Total Suspended Solids

The term turbidity refers to the decreased ability of water to transmit light caused by suspended particulate matter and phytoplankton. Turbidity of pond water has both advantages and disadvantages. In the present investigation the range of turbidity value was recorded as 75 to 110 cms at spot A, 80 to 130 cms at spot B and 74 to 118 cms at spot C.

It is observed that the turbidity was maximum in rainy season in the month of August and minimum value of turbidity was found in summer season in the months of February, April and May at spot A, B and C respectively. The higher sediments during rainy season may inverse turbidity due to the rain fall and mixing the agricultural water, soil with run off rain water. Similar result were observed by Sharma and Gupta (2004), Wagh (1998) and Sinha *et al.* (2004).

The total solids values were found in between the range of 200 to 470 mg/l at Spot A, 210 to 550 mg/l, at Spot B, and 250 to 590 mg/l at Spot C. It was maximum in the month of August at Spot A and Spot B But in the month of September Spot C while it was minimum in month of December at Three spots similar result was observed by Dasgupta and Purohit (2003). Indirabai *et al.* (2005) and Pawar *et al.* (2006). The higher value of total solids during monsoon in the month of August and September may be attributed to an rain fall and increased load of soluble salt from the catchments areas due to the surface run-off.

The total dissolved solids value were found to be in the range between 100 to 365 mg/lit at spot A. 105 to 415 mg/l at Spot B and 145 to 420 mg/l at Spot C. It was found to be maximum value in the month of August at spot A while at spot B and C in the month of September. While minimum value was recorded in the month of January at spot B and C but Spot A in the month of December. It was found the low value in the winter season and high value in the rainy season. It is slightly fluctuated in the summer season due to the leaching of surrounding rain water.

The total suspended values were found to be in the range between 95 to 165 mg/l at spot A, 110 to 180 mg/l at spot B and 105 to 170 mg/l at spot C. It was minimum during winter season and maximum values during rainy season. The maximum value during monsoon might be due to surface runoff.

Acknowledgement

The authors are thankful to Principal, Dr. R. L. Kawale and Dr. D. G. Solunke, Head, Department of Zoology and Fishery Science, Rajarshi Shahu College, Latur for providing necessary facilities.

References

APHA, 1995. *Standard Methods for Examination of Water and Wastewater*, 19th edn. American public Health Origination, American Water Works Association/Water Pollution Control Federation, Washington D.C.

Dasgupta Mahuya Adak and Purohit, 2003. Assessment of the water quality in Rajgangpur industrial complex Part– I: Physico-chemical parameters and water quality index. *Poll. Res.*, 22(1): 133–138.

Goltman, F.L., 1975. *River Ecology*, (Ed.) B.A. Whittom. Blackwell Scientific Publication, Oxford London, Edinburg Malburne, p. 39–80

Hutchison, G.E., 1975. *A Treatise on Limnology, Vol. 1, Part 1: Geography and Physics of Lakes*. John Wiley and Sons Inc., New York, London, p. 540

Indrabai,W.P.S., Nalini, N. and Beebijohn, 2005. Study on water quality in some selected areas of Tiruchirapplli city after the failure of North-East monsoon. *Poll. Res.*, 24(1): 169–174.

Katariya, H.C., 1994. An evaluation of waters quality of Kaliasot river. *IJEP*, 14(9): 690–694.

Paka, Swarnalatha and Rao, Naarsingh, 1997. Inter-relationship of physico-chemical factors of a pond. *J. Eco. Biol.*, 8 (10): 67–72.

Pawar, S.K., Madlapure, V.R., Tale, R.H. and Pulle, J.S., 2006. Seasonal variation in water transparency, total solids, total dissolved solids, total suspended solids of Sirur dam water, Nanded, India. *Ecol. Env. and Cons.*, 12(1): 171–174.

Prasad, B.M. and Sexena, Manjula, 1980. Ecological study of blue green algae in river Gomati, India. *J. Enviro. Hlth.*, 22(2): 151–168.

Reginaa, B. and Nabi, B., 2004. Physico-chemical characterization of Cauvery and Bhavani river at the confluence point Kooduthurai river. *Eco. Env. and Con.*, 10(4): 541–543.

Sharma, R.D., Neerulal and Pathak, D.D., 1981. Water quality of sewage drains entering Yamuna at Agra. *Indian. J. Env. Hlth.*, 23(2): 118–122.

Sinha, D.K., Sexana, Shilpi and Saxena, Ritesh, 2004. Water quality index of Ram Ganga river water at Moradabad. *Poll. Res.*, 23(3): 327–531.

Wagh, N.S., 1998. Hydrobiological parameters of Harsul dam in relation to pollution. *Ph.D. Thesis*, Dr. B.A.M.U., Aurangabad.

Wetzal, R.G. and Likes, G.E., 1979. *Limnological Analysis*. W.B. Saunder Co., Philadelphia, p. 357.

Chapter 8

Study of Physical Parameters of Ujani Reservoir in Solapur District of Maharashtra

☆ *A.K. Kumbhar, D.A. Kulkarni, P.S. Salunke and B.N. Ghorpade*

Introduction

Indian is bestowed with natural resources in the form of extensive coastlines, river systems, estuaries, ponds, tanks, lakes, reservoirs and beels.The hydraulic structure across a river to store water on its up-stream side, which is an impervious or fairly impervious barrier across a natural stream so that a reservoir is formed. This water is then used as and when it is needed. Due to construction of reservoir, water level in that river at its upstream side is very much increased, and a large area may be submerged depending upon the water spread of the reservoir so formed. Due to increased demands for reliable supplies of electric power, irrigation, and drinking water, the number of new hydropower reservoirs is increasing dramatically, especially in India.Ujani reservoir in Solapur district of Maharashtra is a large reservoir, which fulfills the needs of the surrounding rural and urban areas in many ways. In India total reservoir area is 3 million hectare (Sugunan, 1995).

Ujani reservoir is one of the large reservoir in Maharashtra state. The reservoir is constructed across the river Bhima at Ujani village in Madha tahasil of Solapur district of Maharashtra located in 180-04′-0′′N latitude and 750-7′-0′′E longitude. The knowledge of a reservoir ecosystem is of considerable value in assessing the ecological nature of the reservoir, which can be studied by the assessment of the physical characters of the reservoir water. Hence the present study is undertaken to analyze the physical nature of the reservoir.

To face the water scarcity, the Ujani dam was constructed on the Bhima river. The Govt. of Maharashtra was started the construction of the dam in 1964 and it was completed in June 1980. The actual construction work was started in 1968 and it was completed in 1980. The Ujani dam is constructed on the Bhima river, the chief river of Pune district. The Bhima rises on the crest of the Sahyadri near the famous temple of Jyotirling Bhimashankar about 80 km south-east to Mumbai. The main tributaries of the Bhima river are the Velana and Ghod rivers on the left and the Bhama, the Indrayani, the Mula-Mutha and the Nira rivers on the right. Many dams are in place across most of these tributaries. The dams on the Mula-Mutha river and the Nira river are over 100 years old. Some of the other dams have been constructed in the last 25 years.

The Bhima river collects the outflow of several dams built in its upstream. The major dams on the tributaries of Bhima river are Panshet and Varasgao on the Mula-Mutha river, Ghod on the Ghod river, Pawona on Pawona river, Mula on the Mula river and Bhatghar and Veer on the Nira river. The catchmets of the Mula-Mutha, the Ghod and the Mula rivers are densely populated. Moreover, several large and medium agro-industrial plants are concentrated in Pune and Ahmednagar districts of the catchment area. The river collects untreated effluents from these urban centers and treated and untreated effluents from industry. Non-point run off from agricultural fields growing irrigated cash crops is also collected by the Bhima river as it enters in the Ujani reservoir.

51 villages were submerged due to the construction of the Ujani dam. They includes 23 villages from Madha tahashil of Solapur district, 25 villages from Indapur tahasil of Pune district and 3 villages from Karjat tahasil Ahmednagar district.

The agricultural land of 82 villages was submerged due to the construction of the Ujani dam. It includes 30 villages from Madha and Karmala tahasils of Solapur district, 38 villages from Indapur and Daund tahasils of Pune district and 14 villages from Karjat and Shrigonda tahasils of Ahmednagar district.

Table 8.1: The Morphometric Features of Ujani Reservoir

Sl.No.	*Morphometric Features*	*Measurements*
1.	Latitude	18 °–04′–24′′ N
2.	Longitude	75°–07′–15′′ E
3.	Maximum length	2228 m
	Masonary	Non over flow–602 m Over flow–913.40
	Earthern	1626 m
4.	Height	43.285 m
5.	Width	Variable
6.	Maximum depth	40 m
7.	Free catchment area	14850 km^2
8.	Total storage capacity	117.25 TMC
	Live	53.71 TMC
	Dead	63.54 TMC
10.	Basin Slope	Gentle
11.	Submerged area	29000 hectare
12.	Reservoir bottom	Silted
13.	Average rainfall	500 mm

A man made lake, the Ujani reservoir is to serve definite purposes. The main purpose of Ujani reservoir is to provide water for irrigation in Solapur district. The water of the reservoir is also used for drinking purpose to Solapur, Pandharpur, Sangola, Mangalwedha and Akkalkot cities. Many villages located on the bank of the reservoir and Bhima river are also getting drinking water from Ujani reservoir. The Ujani reservoir also provides a very good site for the rearing of commercially important inland fishery. The reservoir also provides natural habitat to plants, birds, animals and that also attract visitors who come for recreation, and indirectly the Ujani reservoir adds diversity to the landscape.

The climate over the entire catchment area varies from moist tropical in the source region of the main river and its tributaries to dry tropical in the immediate vicinity of dam. The average annual rainfall in the source region is about 5000 mm in Sahhyadri hill region and about 500 mm near the Ujani project. The vegetation in the study area is decidous and thorny bush *i.e.,* xerophytic type.

Materials and Methods

The investigation was carried out for two years *i.e.,* from July 2003 to June 2005.The water samples were examined in morning hours.Parametrs like air temperature, water temperature, pH, and transparency were analyzed at the study sites.

The methods used for the analysis of various physical parameters are as given in methodology for water analysis (Kodarkar *et al.,* 1998, and Trivedy and Goel, 1984).

Figure 8.1: A Panoramic View of Ujani Reservoir

Figure 8.2: Collection of Water Sample from Ujani Reservoir

Results and Discussion

Range of physical characters for 24 months period is shown in Table 8.1.Air temperature ranged from 21.2 to 44.1ºC.Water temperature in the reservoir ranged from 19 to 41.8ºC.Water temperature influence the aquatic life and concentration of dissolved gases like CO_2,O_2 and chemical solutes. Changes in the temperature produce characteristic pattern of circulation and stratification (Kulshrestha *et al.*, 1992).The air and water temperatures depend on geographical location and meteorological conditions such as rainfall,humidity,cloud cover, wind velocity at a particular place.Generally,the rise in temperature accelerates the rate of chemical reactions, reduces the solubility of gases, amplifies taste and odour and elevates overall metabolic rate. Reservoir having water temperature more than 22ºC are the highly productive reservoirs (Jhingran and Sugunan 1990, Sugunan 1995).The average water temperature was recorded at 24 and 24.5ºC during first and second year of the study, which reveal that Ujani reservoir is highly productive. The present finding also corroborate with Sakhare (2007), who mentioned Yedari reservoir in Maharashtra as highly productive reservoir. The highest water temperature recorded in summer months can be attributed to the direct relationship between bright sunshine, its duration and air temperature (Laxminarayana, 1965).

Table 8.1: Physical characteristics of Ujani reservoir in Maharashtra

Sl.No.	Parameters	Range
1.	Air temperature (°C)	21.2 to 44.1
2.	Water temperature (°C)	19 to 41.8
3.	pH	7.2 to 9
4.	Transparency (cm)	30.2 to 51.7

pH is one of the most important characteristics regulating the life process and nutrient availability in any waterbody.The pH of water has a significant role in the survival of aquatic plants (Sculthrope, 1967). The pH of water varied from 7.2 to 9, which is in agreement with earlier work (Sinha and Jha, 1997).Generally the pH of water was lower during the period from July to September. The pH range of 6 to 9 is most suitable for pond fish culture, while pH more than 9 is unsuitable (Swingle, 1967).Thus, the pH range indicate that Ujani reservoir is having a pH range of 7.2 to 9 is suitable for survival of fish. Maximum values during summer were probably due to increased photosynthesis in the algal blooms resulting into the precipitation of carbonates of calcium and magnesium from bicarbonates causing higher alkalinity (Kulshrestha *et al.*, 1992).

Water transparency is a physical measurable variable and is quite significant for production. Transparency is inversely proportional to turbidity created by suspended inorganic and organic matter. During the present two year investigation, transparency expressed in cms ranged between 30.2 to 51.7.The water was less transparent during monsoon as compared to winter or summer. Present finding corroborate with finding of Sakhare (2007).

Acknowledgements

Authors are grateful to Dr.Ashok D.Mohekar,Principal,SMDM Mahavidyalaya,Kallam for providing the laboratory facilities to carry out this research work.

References

Jhingran, A.G. and Sugunan, V.V., 1990. General guidelines and planning citeria for small reservoir fisheries management. p. 18. In: *Reservoir Fisheries in India. Proc. of the Nat. Workshop on Reservoir Fisheries*, 3–4 January, (Eds.) A.G. Jhingran and V.K. Unnithan. Special Publication 3, Asian Fisheries Society, Indian Branch, Mangalore, India.

Kodarkar, M.S., Diwan, A.D., Murugan, N., Kulkarni, K.M. and Ramesh, Anuradha, 1998. *Methodology for Water Analysis: Physico-chemical, Biological and Microbiological*. Indian Association of Aquatic Biologists, Hyderabad.

Kulshrestha, S.K.,George, M.P., Saxena Rashmi, Johri, Malini and Shrivsatava, Manish, 1992. Seasonal variation in the limnochemical characteristics of Mansarovar reservoir of Bhopal. In: *Aquatic Ecology*, (Eds.) S.R.Mishra and D.N. Saksena. Ashish Publishing House, New Delhi, pp. 275–292.

Laxminaryana, J.S., 1965. Studies on the phytoplankton of the river Ganges, Varanasi, India, Part I–IV. *Hydrobiol.*, 25(91/2): 119–175.

Sakhare, V.B., 2007. *Reservoir Fisheries and Limnology*. Narendra Publishing House, Delhi.

Sculthrope, C.D., 1967. *The Biology of Aquatic Vascular Plants*. St. Martins Press, New York, 610 pp.

Sugunan, V.V., 1995. *Reservoir Fisheries of India*. FAO Fisheries Technical Report 345, Daya Publishing House, Delhi.

Swingle, H.S., 1967. Standardization of chemical analysis of water and pond muds. FAO Fish. Rep., 44: 397–342.

Trivedy, R.K. and Goel, P.K., 1984. *Chemical and Biological Methods for Water Pollution Studies*. Environmental Publication, Karad.

Chapter 9

Studies on Water Quality Parameters and Prawn Farming of Bastawade Pond from Tasgaon Tahsil of Sangli District, Maharashtra

☆ *S.A. Khabade and M.B. Mule*

Introduction

Water quality plays an important role in the growth of aquatic animals. There should be optimum range for the normal growth of aquatic animals. The water quality standards below and above the optimum range may lead to either stress or death among the aquatic animals. Mainly, the water quality depends on physical, chemical and microbiological parameters. Physico-chemical and microbial characteristics of water body depend upon many factors such as location, dispersal of domestic and industrial waste water and many other activities (Sathe *et al.*, 2000). Various water quality standards are recommended by World Health Organization (WHO), Indian Council of Medical Research (ICMR), Indian Standard Institute (IS) and National Environmental Engineering Research Institute (NEERI). For the normal growth of the freshwater prawns, the optimum range of water quality is also essential.

There are about 100 species of fresh water prawn *Macrobrachium*, identified from various parts of the world of which about 25 species are found in India. Among the Indian prawns, *Macrobrachium rosenbergii, M. malcolmsonii, M. brimanicum, M. idella, M. rudi, M. equidens, M. seacriculum* and *M. lamarrei* are suitable for farming in India. In India, freshwater prawn farming is a growing industry. Among the different species of freshwater prawn, the giant fresh water prawn, *Macrobrachium rosenbergii* (also known as "Scampi" in the trade) is the most commercially important species because of its ability to grow both in fresh water and low saline waters, compatibility for polyculture with carps, omnivorous feeding habits, hardihood and potential to grow fast (Dholakia, 2004). It is seen in almost all major rivers and lakes in India which are connected to sea. It does not have much of problems of different disease attacks and has good consumer preference. It has ever increasing demand both in the domestic as well as export market. Prawn farming is a profitable venture in terms of export value.

Gaint freshwater prawn is a native of India, so that it makes the farming operation easier. In India, there are about 18 million acres of freshwater area suitable for freshwater prawn farming. Mostly, prawn can be grown with other species of fish which is known as polycultre. Prawn farming in ponds is the most technically and economically viable farming method. Eventhough fresh water prawn production from unit are in unit time is less compared to fish, the high market price for prawns makes prawn farming more attractive than fish farming.

The fishery department of Sangli district has grown the prawns in 'Bastawade shallow water pond'. As there is no sufficient baseline data about water quality parameters of the pond under study, hence the present work has undertaken to monitor the water quality parameters, which helps to decide suitability of water for prawn farming.

Study Area

The Bastawade pond is located in Tasgaon tahsil of Sangli district in Maharashtra. The pond is situated in the eastern draught prone hilly area of Tasgaon tahsil, in between Lat. 16°46′ to 17°1′ N and Log 73°42′ to 75°4′ E.

The pond is small sized man-made water reservoir mainly constructed for the drinking water supply to cattles. No data is available regarding the catchment area of the pond.

Material and Methods

Monthly collection of water sample was done from the period July 2000 to April 2001 by using samplers (Plastic containers of 5 litre size). The site selected for water sampling lies near the dams earthen embankment.

The physico-chemical parameters such as water temperature, pH and dissolved oxygen were determined in the field during monthly visits. The total alkalinity, acidity, hardness, magnesium, calcium, chlorides, residual chlorine, free CO_2, hydrogen sulphide, nitrate and phosphate were analysed in laboratory by preserving the water with preservative. The analysis of physico-chemical parameters was done according to the standard method of Trivedy and Goel (1984), APHA, AWWA and WPCF (1985).

Results and Discussion

There is no substitute for good water quality and quantity for fresh water prawn farming. The physico-chemical nature of water and its seasonal variation should be studied at the time of site selection for prawn farming.

Fourteen water quality parameters of Bastawade pond were determined regularly in each month and results were summarized in the table. During months of June 2000 and May 2001, there is no availability of water in the pond, so that there is no analysis of physico-chemical parameters, in these two months.

The physico-chemical characteristics of water recorded are affected by rainfall, temperature, availability of light.

Table 9.1: Water Quality Parameters of Bastawade Pond from July 2000 to April 2001

Parameters	*Months*									
	July 2000	*Aug 2000*	*Sept 2000*	*Oct 2000*	*Nov 2000*	*Dec 2000*	*Jan 2001*	*Feb 2001*	*Mar 2001*	*April 2001*
Water temperature °C	26.0	27.0	28.0	27.6	26.0	25.0	27.0	28.0	31.0	34.0
pH	7.8	8.0	8.2	8.5	8.4	8.3	7.7	7.2	7.1	7.4
Total alkalinity	220	265	150	195	220	220	245	195	200	155
Acidity	42.5	60.0	22.5	20.0	32.5	37.5	30.0	27.5	30.0	20.0
Hardness	216	202	126	122	110	134	140	184	116	200
Magnesium	93.62	94.94	13.18	35.60	5.27	39.55	35.60	14.50	10.54	77.79
Calcium	29.65	23.24	30.46	27.25	51.30	29.65	34.46	24.84	56.11	32.86
Chlorides	42.6	53.96	32.66	42.6	48.28	39.76	42.6	32.66	34.08	137.74
Residual Chlorine	Ab	Ab	Ab	Ab	Ab	Ab	Ab	Ab	Ab	Ab
Dissolved Oxygen (D.O.)	2.41	3.82	4.63	7.04	6.64	7.65	7.44	8.65	5.83	4.02
Free CO_2	17.6	11.0	Ab	Ab	Ab	Ab	Ab	31.0	19.8	11.0
Hydrogen Shulphide (H_2S)	0.14	0.14	0.28	0.425	0.85	0.425	0.566	0.708	3.11	0.85
Nitrate	3.9	4.3	14.4	2.3	0.8	0.3	1.3	1.6	1.8	4.6
Phosphate	0.08	0.01	1.0	0.3	0.01	1.2	0.6	0.6	2.6	2.0

All values are mean of four readings.

All values expressed in mg/L except water temperature and water pH.

In present investigation the water temperature ranged between 25°C to 34°C and shows seasonal variation. Basha and Qureshi (1993) observed that the temperature ranging between 24-28°C was conductive for the abundant occurrence of prawns. The optimum range of temperature for best growth of prawn is about 28°C to 31°C (MPEDA 1994). Gopinathan *et al.* (1978) have reported direct relationship between temperature and number of juveniles of Brackish water prawns.

pH depends on quality of water, photosynthesis, level of ammonia, metabolic wastes etc. The pH of Bastawade pond was alkaline and values ranged from 7.1 to 8.5, being lower in summer and higher in winters. Higher the pH, higher will be the chances for ammonia hazard. In the present study the pH was always found on alkaline side which is considered favourable for aquatic animals like prawns. The lower pH values in summer are observed as a result of decomposition of organic matter at higher temperature, resulting into surplus amount of free carbondioxide (Saha and Pandit, 1986).

Total alkalinity of Bastawade pond ranged between 150 to 265 mg/L. The maximum values were recorded in the month of August 2000 due to less rainfall and minimum values were recorded in the month of September 2000, due to high rainfall. The alkalinity being due to bicarbonates (Misra *et al.*, 1975 and 1976; Bohra, 1976; Bohra *et al.*, 1979). The accumulation of bicarbonate in summer may be due to increased rate of decomposition (Munawar 1970). Basha and Qureshi (1993) have also reported the significant direct correlation of prawn population with alkalinity and total hardness.

Acidity in natural unpolluted freshwaters is mostly due to the presence of free CO_2 in the form of carbonic acid, Trivedy and Goel (1984). In Bastawade pond the acidity reported was ranged between 20.0 to 60.0 mg/L. The abundant occurrence of prawns matched with the peak of alkalinity and total hardness (Manohar and Qureshi 1994).

Hardness is generally caused by the calcium and magnesium ions present in water. The values of hardness ranged between 110 to 216 mg/L. Hardness between 50 to 100 mg/L is permissible for prawn growth and it depends on the basic source of water (MPEDA 1994). Liming raises total hardness.

Magnesium and Calcium have been shown to be increased by sewage pollution (Beeton, 1969; Munawar, 1979). In Bastawade pond the minimum amount of magnesium found was 5.27 mg/L, reported during November 2000. The higher amount of magnesium found was 94.94 mg/L reported during August 2000. The minimum amount of calcium found in Bastawade pond was 23.24 mg/L and the maximum amount reported was 56.11 mg/L.

Chlorides show a range of 32.66 to 137.74 mg/L. According to Thresh *et. al.* (1944), chlorides are generally considered as indicator of pollution. The prawn population shows inverse relation with chloride (Manohar and Qureshi, 1994). The same observations have been reported by Bash and Qureshi (1993).

The residual chlorine was absent throughout the year of investigation in Bastawade shallow water pond.

The dissolved oxygen is stated to be directly related with the prawn abundance (Broad 1962; Basha and Qureshi 1993). According to S. Manohar and Qureshi (1994), the highest peak of dissolved oxygen found in winter fall coincided with the peak of prawn abundance. However, during the pre-monsoon there is peak of prawn abundance and the dissolved oxygen has been found at moderate level. Exactly, the same results were obtained about dissolved oxygen in Bastawade pond. The peak values of dissolved oxygen reported in pond under study are 7.65, 7.44 and 8.65 mg/L found during winter months. However, during pre-monsoon months the dissolved oxygen reported was found to be at moderate level, about 5.83 and 4.02 mg/L.

Qureshi (1994) stated that the occurrence of prawns generally co-incided with the lowest ebb of free carbon dioxide, particularly winter when the free carbon dioxide was found completely absent. However, the pre-monsoon peak of the prawns is observed when the free carbondioxide was present. In the present investigation also, free CO_2 was absent during winter and present during pre-monsoon period and during rainy period. The most of the growth of prawns in Bastawade pond takes place during winter and later few months of summer.

In water reservoirs all the sulphur from organic matter is released in the form of H_2S and later, may get converted into elemental sulphur or sulphate depending upon pH and oxygen regime. Sulphur can be considered as an indicator of organic pollution and prevailing oxygen deficient conditions (Ruttener 1953; Vamos, 1964).

Presence of H_2S in water body is an indication of organic pollution load in it. (Olsen and Sommerfeld, 1977). In the present investigation the concentration of H_2S in the Bastawade pond is not much higher which indicate that there is no high level of organic pollution. The H_2S range found was 0.14 mg/L to 3.11 mg/L which is not harmful for prawn growth.

The nitrates remain much variable throughout the year. According to Ganapati (1960), the non-polluted tropical waters are generally deficient in nitrate but the factors like discharge of sewage, surface runoff and nitrogen fixation may increase its concentration in the water body. In present investigation it has been observed that whatever nitrate present during the year of investigation is due to the surface runoff and nitrogen fixation and not due to the discharge of sewage. Nitrate ranged from 0.3 mg/L to 14.4 mg/L.

The values of phosphates in the present investigation ranged between 0.01 mg/L to 2.6 mg/L. High values of phosphates usually found in reservoirs affecting by domestic sewage (Anantharaj *et al.*, 1987). In Bastawade pond there is no any type of domestic sewage pollution, so that the values of phosphates recorded were not higher but in optimum range. High phosphate may lead to significant, undesirable growth of plankton.

The optimum range of nitrates and phosphates recorded not affecting the growth of the prawns in Bastawade pond.

In the present study, the prawn species *Macrobrachium rosenbergii* which is cultivated in the man made water reservoir, the Bastawade pond shows the luxuriant growth rate. The average weight of the prawns recorded was 100 to 130 gms in 6 to 7 months. The standard period of good growth of prawns is about 8 months to 12 months. Within 5 months the prawns grows upto 50 to 100 gms in weight.

In Bastawade shallow water pond, the post larvae of prawns were deposited during October 2000 and collected the well grown prawns during April 2001. Thus within 7 months, there is a good growth of the prawns, *M. rosenbergii.* It indicates that the water quality (physico-chemical and biological) of Bastawade shallow water pond is good for the growth of the prawns and for fresh water prawn farming in future. According to Saumen Chakrabarti (2005), abundance of *Macrobrachium dayanum* depended upon ecological balance between different biotic and abiotic (physico-chemical) factors. According to Qureshi (1994) the lower abundance of prawns in the vegetation free area was due to the sunlight factor. The water reservoir under study was vegetation free and oligotrophic in nature.

References

Ananthraj, B.V., Bhagyalakshmi and Lakshmi, R., 1987. Limnology of river Cooum with special reference to sewage and heavy metal pollution. *Proc. Indian Acad. Sci. (Anim. Sci.)*, 96: 141–149.

APHA, AWWA, WPCF, 1985. *Standard Methods for the Examination of Water and Wastewater*, 20th Edn. American Public Health Association, Washington DC.

Basha, S.M. and Qureshi, T.A., 1993. Hydrobiological characteristics and prawn population of Halali reservoir of Madhya Pradesh, India. *J. Applied Pure Biol.*, 8: 55–62.

Beeton, A.M., 1969. Changes in the environment and biota of the great lakes. In: *Eutrophication: Causes, Consequences and Correctives.* Nat. Acad. Sci., Washington, DC, pp. 150–187.

Bohra, O.P., 1976. Some aspects of limnology of Padamsagar and Ranisagar, Jodhpur. *Ph.D. Thesis,* Jodhpur University, Jodhpur.

Bohra, O.P., Misra, S.D. and Bhargava, S.C., 1979.Diurnal variation studies on Nakhi lake, Mount Abu, India. *Bioresearch,* 3 : 33–43.

Broad, A.C., 1962. *Environmental Requirement of Shrimp.* Publs. Publi. Hlth. Serv. Washington 1.

Dholakia, A.D., 2004. *Fisheries and Aquatic Resources of India.* Daya Publishing House, Delhi, p. 185.

Ganapati, S.V., 1960. Ecology of tropical waters. In: *Proceedings of Symposium on Algology,* (Eds.) D. Raghavan and P. Kachrro. ICAR, New Delhi, pp. 204–218.

Gopinathan, K., Kaliyanmurthy, M. and Rao, J., 1978. Studies on some species of post-larval peneids of lake Pulicat in relation to their environmental parameters. In: *Proc. Nat. Acad. Sci. Ind.*, 14 : 195–209.

Manohar, S. and Qureshi, T.A., 1994. *Populations of prawns in relation to the Ecological characteristics of River Betwa at Bhojpur*, Madhya Pradesh, India. Palani Paramount Publications ISSN : 0970–937–0006–241.

Misra, S.D., Bhargava, S.C. and Bohra, O.P., 1975. Diurnal variation in physico-chemical factors at Padamsagar reservoir during pre-monsoon period of the year 1974. *Geobios*, 2: 32–33.

MPEDA, 1994. *Freshwater Prawn Farming.*

Munawar, M., 1979. Limnological studies on freshwater ponds of Hyderabad, India II : The biocenose. Distribution of unicellular and colonial phytoplankton in polluted and unpolluted environments *Hydrobiologia*, 36(1): 105–128.

Olsen, R.D. and Sommerfeld, M.R., 1977. The physico-chemical limnology of a desert reservoir. *Hydrobiologia*, 55: 117–129.

Qureshi, T.A., 1994. Results of studies on freshwater prawn carried out in Madhya Pradesh. AFSIB. Special Publ. No. 10 Mangalore, pp. 43–48.

Ruttener, F., 1953. *Fundamentals of Limnology*. University of Toranto press. 1976: Diurnal variation in certain hydrobiological factors and phytoplankton pigments at Padamsagar reservoir. Jodhpur (Raj.). *Trans. Ind. Soc. Desert Tech.*, 1: 18–19.

Saha, L.C. and Pandit, B., 1986. Lentic and lotic systems a review. *Acta Ecologica*, 8: 1–12.

Sathe, Sanjay S., Khabade, S.A. and Hujare, M.S., 2000. Studies on wetlands of Tasgaon tahsil and its importance in relation to fisheries and agricultural productivity, Project Report submitted to University Grants Commission, New Delhi, Western Regional Office, Pune.

Thresh, J.C., Suckling, E.V. and Beale, J.E., 1944. *The Examination of Water and Water Supplies.* London.

Trivedy, R.K. and Goel, P.K., 1984. *Practical Methods in Ecology and Environmental Science.* Enviro Media Publications, Karad (India).

Vamos, R., 1964. The release of hydrogen sulphide from mud I. *Soil. Sci.*, 15(1): 103–109.

Chapter 10

Water Quality Management in Freshwater Fish Ponds

☆ *M.M. Girkar, S.S. Todkari, A.T. Tandale and B.S. Chaudhari*

Introduction

Successful aquaculture depends on healthy fish seed, good quality feed and proper water quality management. Poor water quality reduces growth and affects health of fish. Fish diseases usually occur after stress from impaired water quality. Water quality problems may develop suddenly from environmental phenomena such as heavy rains fall, pond overturn and most probably through improper management. Water quality management is very important aspect if water is limited and runoff or other surface water is the principal source of water for aquaculture.

Generally Principal Water Quality Problems Associated with Fish Ponds

1. Excess phytoplankton development
2. Low dissolved oxygen
3. Production of toxic metabolite

Aspects of Excessive Phytoplankton

Phytoplankton often tints the water green, but may also cause the water to appear blue-green, red or brown. If plankton populations are the principal source of turbidity in ponds, population densities may be determined by several methods

1. Secchi disc visibility (transparency)
2. Dissolved oxygen fluctuation
3. Water appearance

Control of Phytoplankton Population

Phytoplankton populations may be controlled by following methods

Flush Pond

Ponds can be flushed of algae by partially draining and refilling with clean water.

Harvest

Some or all of the fish can be removed, but make sure fish are not "off-flavor" from the algae before you harvest, if the fish are to be sold for food.

Chemical Treatment

Phytoplankton populations may be reduced by treating with copper sulphate ($CuSO_4$) or Aquathol to kill algae.

Manual Method

Aquatic Plants removed by hand picking.

Biological Method

In this method grass carps introduced in aquatic ponds.

Low Dissolved Oxygen

Low dissolved oxygen is the most common stress for cultured fish. Infrequent exposure to low oxygen causes a temporary reduction in food consumption by fish. Chronic low levels of D.O. cause reductions in feeding, feed conversion and growth. Low D. O. in culture ponds is often associated with elevated levels of carbon dioxide (CO_2) and unionized ammonia (NH_3), both of which are toxic to fish. Combination of low D.O. with high CO_2 and NH_3 drastically increases the susceptibility of fish to diseases.

Dissolved oxygen requirements vary between species to species. Most warm water fish can withstand low levels of dissolved oxygen for short periods of time, but levels necessary for adequate growth are at least 4.5 to 5.5 ppm.

Causes of Low Dissolved Oxygen

Lethal chronic levels of low oxygen are most common during July-September when fish weights and feed rates are rapidly increasing. High nutrient levels promote phytoplankton growth and oxygen is often nearly depleted by dawn from high rates of respiration.

Low dissolved oxygen may occur suddenly during late summer or early fall when the first cold fronts occur. Strong, cold winds may cause mixing of surface water with oxygen-poor or anoxic bottom water. If large amounts of partially decayed fish waste, fish feed or plant material are present on the pond bottom, an additional oxygen demand will occur when the water layers mix, and oxygen can be totally depleted within 15-30 minutes.

Control

Emergency aeration can be used to increase D.O. Paddlewheel aerators are the most efficient systems currently used. Water exchange and reduction of feeding rates may also be used to increase D.O. Ponds with low oxygen due to phytoplankton mortality can be fertilized with phosphorus to encourage regrowth of the plankton population. Be careful not to over-fertilize. Add no more than 3 ppm of super phosphate (P_2O_5).

Production of Toxic Metabolite

Toxic metabolites are excreted by fish, bacteria and plankton in fish culture systems. Ammonia, nitrite, nitrate, carbon dioxide and hydrogen sulphide are the major metabolites.

Ammonia

Ammonia (NH_3) is passively excreted through fish gills, but toxic NH_3 concentrations may occur in the body when environmental levels of NH_3 are high. Ammonia reacts with water to form the ammonium ion (NH_4), which is relatively non-toxic to fish. Measurements of ammonia in water are the sum of NH_3 and NH_4 and are called total ammonia nitrogen. Levels of 1-2 ppm NH_3 frequently occur in productive fish ponds. The amount of toxic NH_3 total ammonia nitrogen increases with increases in pH and temperature. Ammonia toxicity usually occurs during warm clear afternoons when photosynthesis has removed CO_2 from the water and raised the pH. Ammonia toxicity is most frequently encountered in hauling or holding facilities.

Ammonia toxicity may cause fish to congregate around inflowing water or at the pond edge. Fish may become lethargic and lie on the bottom or swim slowly near the surface or along pond banks. Gills may be red, swollen and flared. Affected fish do not respond to aeration. Ammonia toxicity may be suspected if fish regularly appear ill in the afternoon and apparently recover by dawn. Ammonia toxicity is most common in ponds with less than 50 ppm total alkalinity or less than 50 ppm calcium hardness. Ammonia may be partially removed from water with paddlewheel aerators during warm, sunny weather. Successful methods of ammonia reduction are pond flushing in small ponds or reduction of feeding rates.

Source

Ammonia is a gas primarily released from the fish gills as a metabolic waste from protein breakdown, with some lesser secondary sources such as bacterial action on solid wastes and urea.

Effects

Ammonia tends to block oxygen transfer from the gills to the blood and can cause both immediate and long term gill damage. The mucous producing membranes can be destroyed, reducing both the external slime coat and damaging the internal intestinal surfaces. Fish suffering from ammonia poisoning usually appear sluggish, often at the surface as if gasping for air.

Control

1. Increase aeration to maximum. Add supplemental air if possible.
2. Stop feeding the fish if detected in an established pond, reduce amount fed by half if starting up a new bio-converter/pond.
3. Check an established pond bio-converter for probable clean out requirement.
4. For an ammonia level of 0.1 ppm, conduct a 10 per cent water change out. For a level of 1.0 ppm, conduct a 25 per cent change out. If the tap water has a higher pH than that of the pond then adding the replacement water may make the situation worse. Chemically treat for twice the amount of ammonia measured.
5. Consider transferring fish if the ammonia level reaches 2.5 ppm.
6. If starting up a new bio-converter/pond, discontinue use of any UV Sterilizers, Ozone Generators, and Foam Fractionators (Protein Skimmers).

Retest in 12 to 24 hours.

Under Emergency conditions only, consider chemically lowering the pH one-half unit but not below 6.0.

Nitrite

Nitrite (NO_2) is a product of bacterial reduction of ammonia. Nitrite replaces oxygen in the blood to form methemoglobin. When concentrations of methemoglobin in catfish blood reach 20-30 per cent, the gill filaments become a chocolate brown color and the fish are diagnosed as having 'brown blood' disease. Brown blood often occurs in the fall, even if ponds are not heavily stocked. Behavioral signs of fish with "brown blood" disease are the same as ammonia toxicity. Affected fish do not respond to pond aeration. Nitrite toxicity can be controlled by adding agricultural limestone to soft water or by adding salt to both soft and hard waters. Chloride concentration between 15-30 ppm usually prevent brown blood problems.

Source

Nitrite is produced by the autotrophic Nitrosomonas bacteria combining Oxygen and Ammonia in the bio-converter and to a lesser degree on the walls of the pond. Just as with Ammonia, Nitrite readings may increase with a sudden increase in bio-converter load until the bacterial colony grows to accept the added material. This can happen following the addition of a large number of new fish to a pond or during the spring as the water temperature increases. Fish activity can often increase faster following a temperature increase than the bacterial action does. A bio converter that becomes partially obstructed with waste and/or develops channels through the media may operate at a reduced effectiveness that can also cause the Nitrite levels to increase.

Effects

Nitrite has been termed the invisible killer. The pond water may look great but Nitrite cannot be seen. It can be deadly, particularly to the smaller fish, in concentrations as low as 0.25 ppm. Nitrite damages the nervous system, liver, spleen, and kidneys of the fish. Even lower concentration over extended periods can cause long term damage. Short term, high intensity, "spikes" which often occur during bio-converter startup may go undetected yet cause problems to develop within the fish months later.

Control

1. Increase aeration to maximum. For a Nitrite level of 1 ppm or greater, add supplemental air, if possible.
2. Stop feeding the fish if detected in an established pond, reduce amount being fed by half if starting up a new bio-converter/pond.
3. Discontinue use of any UV Sterilizers, Ozone generators, and Foam Fractionators (Protein Skimmers).
4. For a Nitrite level less than 1 ppm, conduct a 10 per cent water change out and add 1 pound of salt per hundred gallons of changed water.
5. For a level between 1 and 2 ppm, conduct a 25 per cent water change out and add 2 pounds of salt per hundred gallons of changed water.
6. For a level greater than 2 ppm, conduct a 50 per cent water change. Retest and repeat above in 24 hours.

Nitrate

Nitrate, NO_3-N, measured in ppm, is the third and last measurement used to determine the 'health' of the bio-converter. Nitrate is produced by the autotrophic Nitrobacteria combining oxygen and nitrite in the bio-converter and to a lesser degree on the walls of the pond. A zero nitrate reading, combined with a non-zero nitrite reading, indicates the nitrite-nitrate bacterial converter action is not established. Test kits are available with dual droplet or pill form with color charts. The recommended test kit range 0–200 ppm. A Nitrate test kit is considered nice to have but not required for the average pond. In an established pond with part of the routine maintenance including 5 per cent to 10 per cent water change outs every two to four weeks, Nitrate levels will normally stabilize in the 50-100 ppm range. Concentrations from zero to 200 ppm are acceptable.

Carbon Dioxide

Carbon dioxide (CO_2) is a byproduct of aerobic respiration. Carbon dioxide levels in fish culture ponds cycle daily, with highest levels near dawn and lowest levels in mid-afternoon. Fish affected by CO_2 become listless, and then lie motionless on the pond bottom. Affected fish usually recover rapidly after aeration.

Hydrogen Sulphide

Hydrogen sulphide (H_2S) is excreted by bacteria during anaerobic decomposition of waste products on the pond bottom. Any detectable level should be considered detrimental to fish production. Hydrogen sulfide smells like rotten eggs, and hence may be easily detected. Hydrogen sulphide may cause water to appear gray, and then suddenly clear within a few hours, with a black film on the pond bottom and vegetation.

Control of H_2S is achieved primarily by maintaining oxygen at all depths in culture ponds. Water should be aerated, circulated, or maintained at depths. If hydrogen sulphide is detected, water should be carefully removed from the pond bottom. Replace water slowly, and keep temperature differences less than 5°F (3°C). Do not attempt to remove fish by seining. Stop feeding until plankton growth is observed.

Water Quality Monitoring

Water quality parameters may be monitored by various methods. Various instruments and kits are available for each parameter.

Conclusion

Most water quality problems can be avoided by maintaining low stocking density per meter square and low feeding rates. At higher stocking and feeding rates, farmers must be prepared to monitor water quality and to manage any water quality problems encountered.

Water currents from water inlets or wind-driven wave action may provide variation in water quality. Data from water quality analyses should be recorded and stored for future reference to aid in estimating timing of annual dissolved oxygen problems.

References

Bolte, J.P., Nath, S.S. and Ernst, D.H., 1995. *Pond: A Decision Support System for Pond Aquaculture.* Twelfth Annual Administrative Report, PD/A CRSP, Corvallis, OR. pp. 48–67.

Bolte, J.P. and Nath, S.S., 1996. Decision support for pond aquaculture: Parameter estimation techniques. Thirteenth Annual Administrative Report, PD/A CRSP, Corvallis.

Boyd, C.E., 1982. *Water Quality Management for Pond Fish Culture.* Elsevier Scientific Publishing Company, Amsterdam, New York. Pp. 318.

Boyd, C.E., 1990. *Water Quality in Ponds for Aquaculture.* Birmingham Publishing Company, Birmingham, Alabama.

Boyd, C.E., 1998. *Water Quality for Pond Aquaculture.* Research and Development Series No. 43.International Centre for Aquaculture and Aquatic Environments, Alabama Agricultural Experiment Station, Auburn University, Alabama.

Coche, A.G. and Laughlin, T., 1985. *Soil and Freshwater Fish Culture.* FAO Training Series, No. 6.

Hsiech, C.H., Chao, N.H., L.A. De Olivera Games and Liao, I.C., 1989. Culture practices and status of giant freshwater prawn, *Macrobrachium rosenbergii* in Taiwan. Paper presented at the Third Brazilian Shrimp Farming Congress, 15–20 October, Joao Pessa-PB Brazil.

Ingram, B.A., Hawking, J.H. and Shiel, R.J., 1997. Aquatic life in freshwater ponds: A guide to the identification and ecology of life in aquaculture ponds and farm dams in South-Eastern Australia. Co-operative Research Centre for Freshwater Ecology, Albury, NSW, Australia.

Peter Biro, 1995. Management of pond ecosystems and trophic webs. *Aquaculture*, 129(1–4):373–386.

Rao, K.J., 1991. Reproductive biology of the giant freshwater prawn *Macrobrachium rosenbergii* (de Man) Lake Kolleru (Andhra Pradesh), India. *J. of Anim. Sci.*, 61: 780–787.

Rao, K.J. and Tripathy, S.D., 1993. *A Manual of Giant Freshwater Prawn Hatchery.* CIFA Manual Series 2: 50 p.

Rao, K.J., Rangacharyulu, P.V. and Bindu, R.P., 1994. A two phase larval rearing system for *M.rosenbergii.* Paper presented at the workshop on Freshwater Prawn Farming in India, March 1994, CIFE, Mumbai.

Walker, T., 1994. *Pond Water Quality Management: A Farmer's Handbook.* Turtle Press Pty Ltd, Tas, Australia.

Yoo, K.H., and Boyd, C.E., 1993. *Hydrology and Water Supply for Pond Aquaculture.* Chapman and Hall, New York.

Chapter 11

Studies on Phytoplankton Diversity of Vinjasan Lake in Bhadrawati Town of Chandrapur District, Maharashtra

☆ *P.N. Nasare, N.S. Wadhave, N.V. Harney and S.R. Sitre*

Introduction

The Vinjasan Lake is an ancient lake situated at a distance of 4 km. from Bhadrawati town of Chandrapur district in Maharashtra State. This is an ancient lake in Bhadrawati town now in a highly degraded state. This lake is providing bread and butter to Bhoi community which are regularly taking out crop of Shingada from this lake apart from rearing the fishes.

Phytoplanktons are the pioneer of an aquatic food chain. The productivity of an aquatic environment is directly correlated with the density of phytoplankton. The phytoplankton has great significance as they provide the food for the organisms especially the zooplanktons. The physico-chemical factors are directly related with their production. The phytoplankton is the base of most of the lake food webs and fish production is linked to phytoplankton. Moreover number and species of phytoplankton serves to determine the quality of water body.

The production of phytoplankton is directly correlated with phosphate, silicate and nitrogen content of the lake waters. These three elements are essential for the bloom of phytoplanktons and are always inversely proportional in an aquatic environment because the zooplanktons feed on the phytoplanktons. It is the main item of food for many reservoir organisms like fishes, prawns as well as molluscs. Many researchers have published their work on aquatic environment and ecology of phytoplanktons of freshwaters. Some of which includes the work of George (1962), Kamat (1965), Barhate and Tarar (1981),Patil (1995), More and Nandan (2000), Bahura (2001), Borse *et al.* (2003).

Till date no work is reported on this perennial water body of Bhadrawati town by any researcher. Keeping this point in view present work was undertaken.

Materials and Methods

In the present investigation, four sampling sites from four directions of the Vinjasan lake were selected. The Phytoplankton samples were collected directly from the lake water twice a month for the whole year in plastic bottles of 250 ml. capacity. Lugol's iodine is used for preservation. The phyto planktons were counted by adopting Lackey's drop count method. The genera of phytoplankton's were identified with the help of standard available literature such as APHA, AWWA and WPCF (1975), Needham and Needham (1972) and Edmondson (1965).

Results and Discussion

Total 25 phytoplanktons belonging to 5 different groups *viz.*, Cyanophyceae, Chlorophyceae, Euglenophyceae, Bacillariophyceae and Charophyceae represented the phytoplankton diversity in the Vinjasan lake throughout the year. In the present investigation 10 members of Chlorophyceae, 7 members of Cyanophyceae, 4 members of Bacillariophyceae, 2 members of Charophyceae and 2 members of Euglenophyceae were recorded. Details pertaining to occurrence of different algae are presented in Table 11.1.

Table 11.1: Phytoplankton Biodiversity of Vinjasan Lake, Bhadrawati District Chandrapur during Four Seasons of the Year

Taxa	*Summer*			*Monsoon*			*Post-Monsoon*			*Winter*		
Months	*Mar*	*Apr*	*May*	*Jun*	*Jul*	*Aug*	*Sep*	*Oct*	*Nov*	*Dec*	*Jan*	*Feb*
Cyanophyceae												
Nostoc	+	–	+	–	–	+	–	+	+	+	+	+
Oscillatoria	+	–	+	–	–	–	–	–	+	+	+	+
Rivularia	–	–	–	–	–	–	–	–	–	–	+	+
Microcystis	+	+	+	+	–	–	–	+	–	–	+	+
Cylindrospermum	+	+	–	+	–	–	–	–	–	–	+	+
Anabaena	+	–	+	+	–	–	–	–	–	–	+	+
Spirulina	–	–	+	+	–	–	–	+	–	–	+	+
Chlorophyceae												
Volvox	–	+	+	+	–	–	–	–	–	+	–	+
Chlamydmonas	–	+	+	+	–	+	–	–	–	–	–	+
Pediastrum	–	+	+	+	–	–	–	–	–	–	–	+
Oedogonium	–	+	+	+	–	–	–	–	–	–	+	+
Spirogyra	–	+	+	+	–	–	–	–	–	–	–	+
Closterium	+	+	+	+	–	–	–	–	–	–	–	+
Cosmarium	–	+	–	+	–	–	–	–	–	–	–	+
Hydrodictyon	–	+	–	–	–	–	–	–	–	–	–	+
Ankistrodesmus	+	+	–	–	+	+	+	–	–	–	–	+

Contd...

Table 11.1–Contd...

Taxa	Summer			Monsoon			Post-Monsoon			Winter		
Months	Mar	Apr	May	Jun	Jul	Aug	Sep	Oct	Nov	Dec	Jan	Feb
Euglenophyceae												
Euglena	+	+	+	+	+	–	–	–	–	–	–	+
Phacus	+	+	+	+	+	–	–	–	–	–	+	+
Bacillariophyceae												
Cyclotella	–	–	+	–	+	–	+	–	–	–	–	+
Navicula	+	–	+	–	+	–	–	–	–	–	+	+
Nitzschia	–	+	+	+	+	–	–	–	–	–	–	+
Pinnuularia	–	–	+	–	–	–	–	–	–	–	–	+
Charophyceae												
Nitella	+	+	+	+	–	–	–	–	+	+	+	+
Chara	–	+	+	+	–	–	–	–	+	–	+	+

+: Present; –: Absent.

During early months of study when the temperature was made moderate to some extent, blue greens and greens dominated while as the summer advanced and temperature of water increased the diatoms become dominant. The occurrence of *Microcystis* (the toxin producing blue green algae) (Harris and James 1974) in blooms is a significant feature of tropical waters (Wetzel, 1975). In the present study also this alga appeared during summer months in dominance. Similar blooms have also been reported earlier in waters of Western Rajasthan by Bohra (1976) and Misra *et al.* (1978). The high water temperature and low dissolved oxygen during summer create favourable conditions for the development of blue green algae (Fritsch, 1907). According to Ganpati (1940) and Chakrabarty *et al.* (1959) abundance of blue greens is associated with high temperature and low pH.

Total algae were less in August months due to low pH, Low D.O., turbid waters and waters diluted by rains. Such observations were also made by Kumar and Singh (2000).

Chlorophyceae were dominant in April months when water was slightly hard, alkaline as well as well oxygenated with moderate range of nutrients (Silicate, nitrate and phosphate). Similar observations were made by Gonzalves and Joshi (1946), George (1962) and Chakrabarty *et al.* (1959). Algae were abundant in February (Winter) months when pH, D.O., Calcium, COD and EC were maximum and total hardness, chloride, silicate, nitrate and water temperature were minimum whereas TS, TDS and Phosphate were moderate. George (1962), Kumar (1994) and Kumar and Singh (2000) suggested that high pH,DO, COD and EC promotes the growth of Phytoplanktons.

Diatoms (Bacillariophyceae) were maximum in May. Similar observations were made by Hutchinson *et al.* (1929) and Somshekhar (1987).

The present investigation clearly reveals that the lake ecosystem is rich and fertile and supports diverse range of phytoplanktons throughout the year.

Acknowledgements

The authors are thankful to Shri Nilkanthrao Shinde Ex MLA and Secretary Bhadravati Shikshan Sanstha, Bhadrawati for providing laboratory facilities to carry out the research work and to

Dr.K.D.Thengane, Principal, N.S. Science and Arts College, Bhadrawati, District Chandrapur for encouraging to undertake the research work and providing constructive suggestions.

References

APHA, AWWA and WPCF, 1975. *Standard Methods for the Examination of Water and Wastewater*, 14th Edition, Amrican Public Health Association, New York, 1193 pp.

Barhate, B.P. and Tarar, J.L., 1981. Algal Flora of tapi rivers, Bhusaval, Maharastra. *Phykos*, 20: 75–78.

Bahura, C.K., 2001. Phytoplanktonic community of a highly eutrophicated temple tank, Bikaner, Rajasthan. *J. Aqua. Biol.*, 16(1 and 2): 1–4.

Bohra, O.P., 1976. Some linmonoglcal aspects of Rani Sagar and Padma Sagar (Jodhpur). *Ph.D. Thesis*, Jodhpur University, Rajasthan, India.

Borse, S.K., Lohar, P.S. and Bhave, P.V., 2003. Hydrobiological study of algae and Aner River, Jalgaon, Maharashtra,India. *J. Aqua. Biol.*, 18(1): 15–18.

Chakrabarty, R.D., Roy, P. and Singh, S.B., 1959. A quantitative study of the plankton and the physico-chemical conditions of the river Jamuna at Allahabad in1954–55: *Indian J. Fish*, 6(1): 186–203.

Edmondson, W.T., 1965. *Freshater Biology*. John Wiley and Sons, New York.

Fritsh, F.E., 1907. A sublethal and freshwater algal flora of tropics: A phytogeographical and ecological study. *Ann. Bot.*, 21: 235–275.

Ganpati, S.V., 1940. The ecology of a temple tank containing a permanent blooms of *Microcystis aeruginosa*. *J. Bom. Nat. Hist. Soc.*, 46(1): 155–176.

George, M.G., 1962. Occurrence of a permanent algal bloom in a fish tank at Delhi with special reference to factors responsible for its production. *Proc. Indian Sci. Acad.*, 56: 354–362.

Gonzalves, E.A. and Joshi, B.D., 1946.Freshwater algae near Bombay–1. The seasonal succession of the algae in a tank at Bandra. *J. Bom. Nat. Hist. Soc.*, 46(1): 155–176.

Harris, D.O. and James, D.E., 1974. *Toxic Algae. Carolina Tips*, 37: 13–14.

Hutchinson, A.H., Lucas, C.C. and McPhall, M., 1929. Seasonal variations in the chemical and physical properties of water of the Strait of Georgia in relation to phytoplankton. *Trans. Roy.Soc.Conf.* Series–3, 23 (Part –11), Sect. 5: 177–183.

Kamat, N.D., 1965. Ecological notes on algae of Kolhapur. *J. Biol. Sci.*, 8: 47–54.

Kumar, A., 1994. Seasonal trends in biological and physico-chemical properties of a fish pond at Dumka (Bihar). *Acta. Ecol.*, 16(1): 50–57.

Kumar, A. and Singh, N.K., 2000. Phytoplankters of a pond at Deoghar, India. *Phykos*, 39(1 and 2): 21–33.

Mathur, M. and Pathak, N., 1990. Some Chlorococalles new to India. *Ibid*, 29: 111–113.

More, Y.S. and Nandan, S.N., 2000. Hydrobiolgical study of algae of Panzara river (Maharashtra). *Ecol. Env. and Cons.*, 6(1): 99–103.

Mahajan, S.R. and Nandan, S.N., 2004. Blue green algae of Hartala lake of Jalgaon Maharashtra. *J. Aqua. Biol.*, 19(1): 11–12.

Misra, S.D., Bhargava, S.C., Jhankar, G.R. and Dey, T., 1978.Hydrobiologyuand productivity of some freshwater reservoir and lakes of semi-arid zone, Jodhpur (Rajasthan), UGC Project Report, University of Jodhpur, Jodhpur, India.

Needha, J.G. and Needham, K.R., 1978. *A Guide to the Study of Freshwater Biology*. Holden Day Inc. Pub. San Franscisco, 107 pp.

Patil, Sandhya, 1995. Algal flora of polluted water of Khandesh of Maharashtra. *Ph.D. Thesis,* Poona University.

Somshekhar, R.K., 1987. Periodicity of diatoms and the chloride concentrations in river Cauvery. *J. Indian Bot. Soc.*, 66: 161–165.

Tiwari, A., Upadhaya, R. and Chauhan, S.V.S., 2001. A systematic account of Chlorocoales from Kitham Lake, Agra. *Ibid,* 40(1 and 2): 103–105.

Wetzel, 1975. *Limnology*. W.B. Saunders Co., Philadelphia, 734 pp.

Chapter 12

Aquatic Protected Areas in River Narmada Around Hoshangabad

☆ *Vipin Vyas*

Introduction

Role of protected areas in conservation of biological diversity has been well understood and documented but very little is known about Aquatic Protected Areas even at global level. In India several aquatic protected areas have been notified as intentionally protected areas. Some water bodies fall within land based protected areas like wildlife sanctuaries and national parks and therefore they are incidentally protected areas.

Protected areas can play an important in conservation of freshwater fishes in India but their contribution and present status of fish conservation is still unknown. (Kapoor and Sarkar, 2005).

Fish Biodiversity of River Narmada

Narmada River originating from hills of Amarkantak in district Shahdol of MP has a stretch of 1280 km draining through three states MP, Maharashtra and Gujarat in between Vindhyan and Satpura ranges. About 84 per cent of river stretch (1077 km) falls in Madhya Pradesh covering seven districts. This river takes in its fold 41 principal tributaries on its way and about 25 per cent of the catchment area is covered by the precious tropical rain forests.

The river supports a rich biodiversity of 95 species of fishes as reported by various workers in different investigations since 1941 (Dubey, 1994). Of these, as many as 20 species have been reported to be at risk by several workers mainly due to over exploitation, water pollution and destruction and alteration of habitats.

Table 12.1: Fish Species at Risk in River Narmada

Rao (1992)	12 sps
Desai (1992)	17 sps
CICFRI (Desk Report) 1993	7 sps
Dubey (1994)	20 sps
NBFGR (1995)	9 sps
Arya *et al.* (2001)	15 sps

In 1983 Govt. of MP has made fishing free in rivers. Fishing of brood stocks and juveniles are the main cause of this drastic depletion of fish resources in River Narmada, in addition to alteration of habitats and destruction of breeding grounds due to construction of dams and siltation in river bed (Dubey and Ahmad, 1995).

Aquatic Protected Areas in River Narmada

In order to conserve fish fauna Govt. of Madhya Pradesh has declared several No Fishing Zones in various rivers and streams under MP Riverine Fisheries Rules (Anon, 1971). Out of these 10 areas fall on River Narmada. Recently, 16 deep pools identified on the river have also been included in the list of notified no fishing areas (Anon, 2000). Presently, there are 26 no fishing areas in river Narmada which can be considered as Aquatic Protected Areas. These aquatic protected areas cover 59 km of the entire stretch of 1077 km of the river in MP. Besides these notified areas, there are several *ghats* on the river where fishing is not practiced because of some religious reasons. These can be termed as *sacred ghats* like *sacred groves* in terrestrial areas.

Protected Area Network around Hoshangabad

Hoshangabad, a district headquarters of MP situated on the bank of river Narmada at 682 km downstream from its origin. River Narmada form district boundaries of Hoshangabad, Raisen and Sehore district in this region. As stated earlier there are 8 protected areas fall in this 22 km stretch forming a network of protected areas. Table 12.2 shows aquatic protected areas in this stretch. It indicates that this stretch has 8 protected areas. 7 of them have been noticed by the state govt. under fisheries rules. 3 areas have religious significance and two out of these enjoy both notified status and religious significance.

Table 12.2: Location of Aquatic Protected Areas around Hoshangabad

Category	*Location (District)*	*Stretch*
Notified	Shahganj	½ km upstream and ½ km downstream of Shahganj Guest House
	Joshipur	1 km
	Railway Bridge	½ km
	Bandua	1 km
Notified and Religious	Bandrabhan	1 km
	Budni kund	1 km upstream to 1 km downstream
Religious	Sethani Ghat	1 km

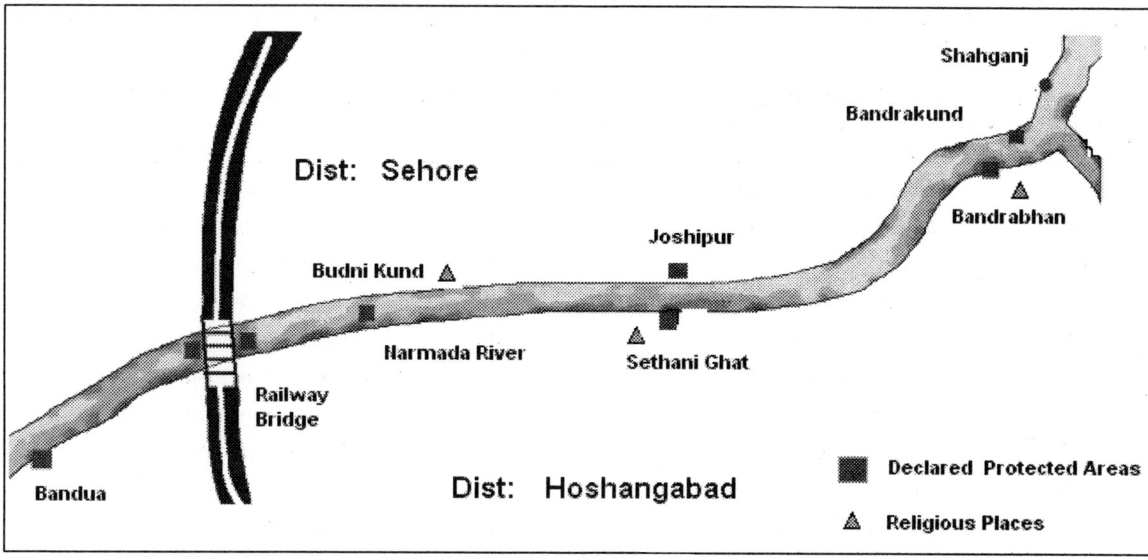

Figure 12.1: Map of Narmada River in Hoshangabad Region Showing Protected Areas

Issues related to Protected Areas

As mentioned earlier, issues and strategies outlined at global level are also relevant in this region River Narmada is also subject to various environmental changes. The legislation which covers the notifications of such closed areas does not have any bearing on the watershed/catchment of the particulars stretch of river. Hydrology of the river has also been altered drastically by construction of dams on main river at Bargi and Punasa. Dams constructed on Tawa, Barna and Kolar, three main tributaries of River Narmada have also constructed to alteration in flow regime of main river.

Socio-economic and cultural aspects have also been considered as key issues in Aquatic Protected Areas. River Narmada supports livelihood and food security at local level, hence community involvement in conservation measures can not be ignored.

Initiative Taken in Madhya Pradesh

Considered the significance of aquatic protected areas in conservation of fish biodiversity and looking to the lacuna in documentation of existing aquatic protected areas the Department of Limnology, Barkatullah University, Bhopal has come up with an initiative to start documentation of these protected areas in MP and has started documentation of conservation status in some aquatic protected area of River Narmada around Hoshangabad along with its socio-economic aspects.

Current Fish Biodiversity Profile of the Area

Several studies have been carried out by various workers on the river Narmada and some of them cover this stretch also. Hora and Nair (1941) reported 40 species of fishes from hill streams of Satpura ranges in Hoshangabad district. Later, Karamchandani (1967) reported 77 species of fishes in a survey conducted during 1959-64. Another survey conducted by the Department of Fisheries (Govt. of MP) in the year 1967-71 covering a larger stretch of the river from Jabalpur to Khalghat reported 46 species. However Vyas *et al.* (2007) reported 47 species in his stretch which indicate richness of this stretch.

Table 12.3: List of Fishes Recorded in River Narmada in Hoshangabad Region

Sl.No.	Name of Fish	Sl.No.	Name of Fish
1.	*Amblypharyngodon mola*	24.	*Mystus aor*
2.	*Anabas testudineus*	25.	*Mystus blekeri*
3.	*Barilius bandelisis*	26.	*Mystus seenghala*
4.	*Barilius barila*	27.	*Mystus tengra*
5.	*Chanda nama*	28.	*Nemacheilus botia*
6.	*Chanda ranga*	29.	*Notopterus notopterus*
7.	*Channa gachua*	30.	*Nandus nandus*
8.	*Channa marulius*	31.	*Ompok bimaculatus*
9.	*Channa striatus*	32.	*Ostobrama cotio*
10.	*Chela laubuca*	33.	*Oxygaster bacaila*
11.	*Clarius batrachus*	34.	*Oxygaster clupeoides*
12.	*Clupisoma garua*	35.	*Oxygaster gora*
13.	*Colisa fasciatus*	36.	*Puntius amphibious*
14.	*Danio devario*	37.	*Puntius chola*
15.	*Glossogobius giuris*	38.	*Puntius chrysopterus*
16.	*Heteropneustes fossilis*	39.	*Puntius conchonius*
17.	*Labeo bata*	40.	*Puntius sarana*
18.	*Labeo gonius*	41.	*Puntius ticto*
19.	*Labeo calbasu*	42.	*Rasbora daniconius*
20.	*Labeo fimbriatus*	43.	*Rita rita*
21.	*Lepidocephalichthys guntea*	44.	*Tor tor*
22.	*Mastacembelus armatus*	45.	*Wallago attu*
23.	*Mastacembelus pancalus*	46.	*Xenentodon cancila*

River stretch in this region is reported to have as many as 6 spots from where seed of mahseer fish are collected in addition to commercial fishes. (Dubey 1984, Langer *et al.*, 1986-87). Recent reports indicate that the quantity of seed at these centers have drastically gone down. (Dutt and Tiwari 2002). One important seed collection center near Bandrabhan (at the confluence point of Tawa and Narmada) has totally vanished due to construction of Tawa dam. (Desai, 1987). This region also has its significance in fisheries as it falls between Bargi and Indirasagar reservoirs constructed on main channel of River Narmada and three main tributaries namely Tawa, Barna and Kolar.

Observations indicate that the protected areas notified by the Government are not very effective as compared to areas where fishing is not performed due to religious reasons. Similar observations were recorded at various other places like in rivers of Himachal Pradesh (Dhanze and Dhanze, 2004) and River Betwa in Madhya Pradesh (Vyas *et al.*, 2008).

Conclusion

It has been experienced that despite of many regulatory instruments and conservation measures, fish population is still declining in rivers. (Dubey and Ahmed 1995). Dehadrai *et al.* (1998) emphasize on holistic approach to enforce various regulations to conserve fish biodiversity. Kapoor *et al.* (1998) and Kapoor and Sarkar (2005) categorically stated the involvement of target community and local self governments in conservation efforts.

In the present case also, we recommend that community participation should be ensured while imposing any legislative measure to conserve fish biodiversity.

References

Anon., 1971. Fisheries Department, M.P. *Fisheries Survey in Narmada River*, 1967–71.

Anon., 1995. *Perspective Plan of NBFGR*, Lucknow (ICAR).

Anon., 2000. Part of Manual on Habitat and Biological Inventory under and NATP Germplasm Inventory under Gene Banking of Freshwater Fishes, NBFGR, Lucknow.

Arya, S.C., Rao, K.S. and Shrivastava, S., 2001. Biodiversity and fishery potential of Narmada Basin Western Zone (M.P., India) with special reference to fish conservation. In: *Environment and Agriculture: Agriculture and Pollution in South Asia*, pp. 168–112.

Central Inland Capture Fisheries Research Institute (Indian Council of Agriculture Research) Barrackpore, 1993, West Bengal, India (A desk review submitted to Narmada Control Authority, Indore).

Dehadrai, P.V., Das, P. and Verma, S.R., 1994. *Threatened Fishes of India*. Natcom. Pub. No. 4: Muzaffarnagar, 412 pp.

Desai, V.R., 1987. Save mahseer: The game fish. *Indian Farming*, 37(4): 33.

Desai, V.R., 1992. Endangered, vulnerable and rare fishes of river system in Madhya Pradesh. In: *Proceeding of the National Seminar on Endangered Fishes of India*, held during 25–26 April at N. B. F. G. R. Allahbad, Abstract No. 22.

Dubey, G.P., 1994. Endangered, vulnerable and rare fishes of West Coast river system of India. In: *Threatened Fishes of India*. Natcon Publication, 4: 77–95.

Dubey, G.P. and Ahmed, 1995. *Problems of Fish Conservation of Freshwater Fish Genetic Resource in India, and Some Possible Solution.*

Dutt, S. and Tiwari, V.K., 2002. Mahseer (*Tor tor*) seed collection centres in and around Hoshangabad. In: *Proc. of National Workshop on Biodiversity and Conservation of Aquaic Resources with Reference to Threatened Fish Mahseer*, p. 117–121.

Dhanze, J.R. and Dhanze, R., 2004. Role of protected habitat in the conservation of indigenous fish germplasm: A case study. In: *Fish Diversity in Protected Habitats*. Nature Conservators Publication, 8: 129–140.

Hora, S.L. and Nair, K.K., 1941. Fishes of Satpura Range, Hoshangabad District, Central Province. *Rec. Indian Mus.*, 43(3): 361–373.

Karamchandani, S.J., Desai, V.R., Pisolkar, M.D. and Bhatnagar, G.K., 1967. Biological investigation on the fish and fisheries of Narmada River (1958–66). *Bull. Cent. Inland Fish. Res. Inst., Barrackpore* (10): 40 p. (Mimeo).

Kapoor, D., Mahanta, P.C. and Pande, A.K., 1998. Ichthyodiversity of India: Status and Conservation. In: *Fish Gen. Biodiversity Conserv.*, Natcon Pub., p. 47–53

Kapoor, D. and Sarkar, U.K., 2005. Priority research components of NBFGR for conservation of Indian fish biodiversity. *Fishing Chimes* 25(1): 110–113.

Langer, R.K., Somalingam, J., Krishna Prasad, J. and Upadhyay, S.K., 1987. Preliminary Observations on the availability of *Tor tor* near Hoshangabad. *J. Indian Fish Association*, 16–17: 63–67.

Rao, K.S., 1991. *Pre- and Post-Limnological Studies on Narmada River.* Final Technical Report submitted to NVDA.

Vyas, V., Vivek, P., Damde, D. and Tuli, R.P., 2007. Mapping of mesohabitat diversity in river Narmada in Hoshangabad region. *Fishing Chimes*, 27(9): 49 –53.

Vyas, V., Vivek, P. and Damde, D., 2008. *Documentation of Aquatic Biodiversity in Rivers and Ponds of MP.* Report submitted to MP State Biodiversity Board.

Chapter 13

Selective Study on the Availability of Indigenous Fish Species having Ornamental Value in Some Districts of West Bengal

☆ *A.K Panigrahi, Sarbani Dutta (Roy) and Indranil Ghosh*

Introduction

Now a day the term Ornamental fish needs no introduction to us. The global trade of this aqua business is increasing rapidly (@ 6 per cent annually; Ghosh, 2005). In aquaculture sector, the ornamental fish breeding, culture and trade provide excellent opportunities as a non–food fishery activity for employment and income generation. It is totally environmental friendly, socially acceptable and involves low investment for adopting as a small scale enterprise with high return. Their attractive coloration, quietness, eco-friendly behaviour and other ornamental peculiarities become the source of joy and peace of people irrespective of age group. But ornamental fish is more a fashion. New fish varieties are seen from time to time. By concentrating on such fish only, we may loose our indigenous stocks some of which have reached towards the door of extinction. Many indigenous ornamental fishes are very much useful for developing new strains to compete in world market. They are used as a tool in biotechnological research in all over the world (Swain *et al.*, 2007).

The history of culturing these fishes in West Bengal is age old. A rich aquatic biodiversity, favourable condition, cheap labour and easy distribution make West Bengal as a pre-eminent hub for this promising industry (Ghosh, 2006). Most of the indigenous and endemic fish species available in this state are highly potential for the purpose of ornamental fish culture. But it is very surprising that severe depletion in the natural fish population of the state occurs due to destruction of habitat, unsustainable mode of exploitation and other stresses. So with the present investigation, an attempt has been made to enlighten the overview of availability of indigenous ornamental fishes in some districts of West Bengal.

Materials and Methods

A survey was carried out for 60 days (September-October,2008) in four districts of West Bengal namely Howrah, Hooghly, North and South 24 Parganas because these districts are rich in different types of indigenous fish species which have got tremendous ornamental status in the international market. A considerable number of fish culture units have been concentrated in these four districts (Ghosh, 2005)

Study Area

The present investigation was carried out in different areas of selected four districts which have been shown in the Table13.1. The entire region under study is highly variable as far as topographic and climatic conditions are concerned. This region enjoys a tropical monsoonal climate, receiving an annual medium range of rainfall with high temperature in summer (30–39°C) and a sharp fall of temperature in winter (15–25°C).

Table 13.1: Selected Areas of the Districts where Survey has been Conducted

Districts	*Name of the areas*
Howrah	Pakuria, Domjur, Kadamtala,Amta, Ramrajatala
Hoogly	Dankuni, Chuchura, Srerampur, Mogra, Bandel, Saorafulli
North 24 pgs	Naihati, Amdanga, Barasat, Barrackpore, Khardah, Baranagar, Bonhooghly
South 24 pgs	Amtala, Mahestala, Falta, Bazbaz, Baruipur, Canning south, Dimond harbour

Selection Criteria of Ornamental Fishes

For the present study, small to medium (<20 cm) sized fish specimens (ideal for aquarium rearing) available in the aquatic habitats of the region were taken into consideration to assess their potentiality for aquarium purpose. Basic parameters considered for the assessment were (*i*) Adult size and hardiness, (*ii*) Attractiveness (colouration pattern, body morphology etc), (*iii*) Ability to thrive in confined environment with supplementary food, (*iv*) Endemicity, (*v*) Behavioural and environmental compatibility with other species.

Identification of the Indigenous Fishes

The indigenous fish fauna having high potentiality as ornamental fishes that are available through out the study areas of the selected districts were identified according to Nelson (1994) and Talwar and Jhingran (1991).

Result and Discussion

During the present investigation 30 species of indigenous fishes with potentiality as aquarium fishes belonging to 22 genera and 13 families have been recorded from the selected areas of the four districts (Table 13.2). Among the families cyprinids showed most dominant family followed by Bagridae, Siluridae, Ambassidae, Notopteridae, Belontiidae, Belonidae, Chacidae, Mastacembalidae, Cobitidae, Nandidae, Anabantidae and Channidae.

Table 13.2: List of Fishes Frequently Available in the Selected Areas

Sl.No.	Local Name	Common Name	Scientific Name	Family
1	Pholui	Black knife fish	*Notopterus notopterus* (Hamilton-Buchanan)	Notopteridae
2	Chital	Humped feather back	*Notopterus chitala* (Hamilton-Buchanan)	Notopteridae
3	Kalbasu	All black shark	*Labeo calbasu* (Hamilton-Buchanan)	Cyprinidae
4	Kanchan pungti	Rosy barb	*Puntius conchonius* (Hamilton-Buchanan)	Cyprinidae
5	Gilli pungti	Golden barb	*Puntius gelius* (Hamilton-Buchanan)	Cyprinidae
6	Pothia pungti	Two spot barb	*Puntius ticto* (Hamilton-Buchanan)	Cyprinidae
7	Sophore pungti	Spot fin swamp barb	*Puntius sophore* (Hamilton-Buchanan)	Cyprinidae
8	Maurala	Mola carplet	*Amblypharyngodon mola* (Hamilton-Buchanan)	Cyprinidae
9	Anju	Zebra danio	*Brachydanio rerio* (Hamilton-Buchanan)	Cyprinidae
10	Dangila danio	Dangila danio	*Danio dangila* (Hamilton-Buchanan)	Cyprinidae
11	Bashpata	Devario danio	*Danio devario* (Hamilton-Buchanan)	Cyprinidae
12	Rasbora	Gangetic scissor tailed	*Rasbora rasbora* (Hamilton-Buchanan)	Cyprinidae
13	Gunte	Guntea loach	*Lepidocephalus guntea* (Hamilton)	Cobitidae
14	Tengara	Golden cat fish	*Mystus tengara* (Hamilton-Buchanan)	Bagridae
15	Tengara	Striped dwarf cat fish	*Mystus vittatus* (Bloch)	Bagridae
16	Aar	Long whiskered cat fish	*Aorichthys aor* (Hamilton-Buchanan)	Bagridae
17	Ritha	Rita	*Rita rita* (Hamilton-Buchanan)	Bagridae
18	Pabdah	Gulper cat fish	*Ompok pabda*(Hamilton)	Siluridae
19	Pungas	Indian tiger shark	*Pungasius pungasius* (Hamilton-Buchanan)	Siluridae
20	Chacca	Indian chaca	*Chaca chaca* (Hamilton-Buchanan)	Chacidae
21	Kankley	Long nosed Needle fish	*Xenontodon cancila* (Hamilton-Buchanan)	Belontiidae
22	Nama chanda	Elongated glass perchlet	*Chanda nama* (Hamilton-Buchanan)	Ambassidae
23	Ranga chanda	Indian glass fish	*Pseudoambasis ranga* (Hamilton-Buchanan)	Ambassidae
24	Nadosh	Leaf fish	*Nandus nandus* (Hamilton-Buchanan)	Nandidae
25	Koi	Climbing perch	*Anabus testudineus (Bloch)*	Anabantidae
26	Khalisa	Stripped gourami	*Colisa fasciata (Schneider)*	Belonidae
27	Khalisa	Dwarf gourami	*Colisa lalia* (Hamilton-Buchanan)	Belontiidae
28	Chuna khalisa	Sunset gourami	*Colisa sota* (Hamilton-Buchanan)	Belontiidae
29	Pankal	Spiny Green eel	*Mastacembalus puncalus* (Hamilton-Buchanan)	Mastacembelidae
30	Lata	Spotted snake head	*Channa punctatus (Bloch)*	Channidae

Observation during the Investigation

Through out the survey it is found that the natural resources of the indigenous fish fauna are in declining trend. The condition may be due to increasing trend of aquatic pollution mainly by agricultural pesticides followed by discharge of pollutants along with industrial effluents from various sources.

It is investigated that the habitats and breeding grounds of these valuable fishes are under threat due to the application of toxicant to prepare the water bodies for the sake of scientific composite fish culture or poly culture after eradicating the weed fishes, now known as indigenous ornamental fishes. The study further revealed that if such condition is allowed to continue then a time will come when these valuable indigenous fish species will not be available to us. So now, time has come to restore all such indigenous fishes and their captive culture is to be popularized. We urge the Government Departments, Private agencies, other Non Government Organizations to boost the culture of such fish species to save the progeny. So in this respect it is very essential to take immediate attention to popularize the captive culture as well as breeding of these indigenous fish species of Bengal.

According to our study it is observed that presently, these resources are exploited in an unorganized way by some unscrupulous traders who operate this trade from outside the region. Their trade usually based on collection of certain target species which have high market value. This mode of exploitation would no doubt contribute largely to the depletion of natural stock of the said resources in the near future. A number of rural people both men and women are engaged in the wild collection and culture of the potential fish resources. But the fact is that the indigenous fishes produced in these areas or wildly collected are mostly meet the demand of local market and the consumers. Lack of awareness and ignorance among rural masses, deficiencies of infrastructure and inadequate policies of the concerned authority have been identified as major hurdles in the way of development of this highly promising sector of commercial fisheries.

Conclusions

Considering the tremendous prospects of these areas it is the prime time to make an integrated effort to promote this sector of fisheries in a sustainable way. Eradication of the ignorance and the enlightenment or awakening of awareness among the rural masses is the essential criteria to restore those valuable fishes.

Following measures have been suggested for sustainable utilization of these potential aquatic resources to enhance the fisheries sector to earn more foreign exchange.

- ☆ Abundant and systematical development of a comprehensive database including feeding, breeding and environmental requirements of each potential indigenous fishes.
- ☆ To evolve sustainable farming technologies for the commercial important native ornamental fishes.
- ☆ Provision of better extension support in the form of technology, finance and marketing to the needy fishermen particularly to the unemployed person to motivate them in this aqua business.
- ☆ Development of captive culture fisheries in 'beels', reservoirs and wet-lands in co-operative basis with some other culture.
- ☆ Establishment of periodic monitoring of the health of the aquatic fauna as well as water quality parameters.
- ☆ Enforcement of fisheries and Environmental Legislations and Acts in the maintenance of resource sustainability and to check unauthorized wild collection of targeted species.

References

Andrews, C., 1992. The ornamental fish trade and fish conservation. *INFOFISH International*, 2: 25–29.

Bassieer, G., 1994. The international trade in aquarium ornamental fish. *INFOFISH International*, 5: 15–17.

Das, J.N. and Biswas, S.P., 2008. Ornamental fish resources of North-Eastern states and their trade potential. *Ornamental Fish Breeding, Farming and Trade*, Kerala, pp. 51–61.

Ghosh, Indranil, 2005. Breeding and larval rearing of ornamental fishes under captive condition with special reference to West Bengal. Course Manual, Regional Training-cum-Workshop on Ornamental Fish (sponsored by MPEDA, Ministry of Commerce, Govt. of India), organized by College of Fisheries, Assam Agricultural University, Raha, Assam, India, Feb. 22–25, pp. 19–23.

Ghosh, Indranil, 2006. The ornamental fish sector in India and the West Bengal scenario. *INFOFISH International*, 4(July/August): 28–31.

Ghosh, Indranil, 2007. Development of fisheries through biotechnology: An appeal. In: *Advances in Aquatic Ecology*, Vol. 1, (Ed.) V.B. Sakhare. Daya Publishing House, Delhi, pp. 95–99.

Mukherjee, M., Dutta, A., Sen. S. and Banerjee, R., 2002. Ornamental fish farming: A new hope in micro-enterprise development in Periurban area of Kolkata, HRD in fisheries and aquaculture for Eastern and North eastern India, CIFE, Kolkata Centre, 14–15 March, pp. 48–59.

Nelson, Joseph S., 2006. *Fishes of the World*. John Wiley and Sons, Inc.

Sahu, B.B. and Mohanty, S., 2000, Quality control of ornamental fish for export compandium of lectures on ornamental fish breeding and culture. *CIFE*, Bhubaneswar, pp. 55–62.

Ramchandran, A., 2002. *Riverine and Reservoir Fisheries of India*, (Eds.) Boopendranzth, *et al.*, pp. 109–134.

Swain, S.K., Mallik, D., Mishra, S., Sarkar, B., and and Routray, P., 2007. Ornamental fish as model animals for biotechnological research. In: *Environmental Biotechnology*, (Eds.) Mishra and Juwarkar. APH Publ. Corp., Delhi, pp. 293–328.

Talwar, P.K. and Jingran, A.G., 1991. *Inland Fishes of India and Adjacent Countries*. Oxford and IBH Pub. Co. Pvt. Ltd., New Delhi, India.

Chapter 14

Utilization of Seaweeds in India and Global Scenario

☆ *G. Thirumaran, R. Arumugam and P. Anantharaman*

Introduction

Seaweeds or marine algae are the primitive type of plants and they are growing abundantly in the shallow waters of sea, estuaries and backwaters. They flourish wherever rocky, coral or suitable substrata are available for their attachment. They belong to three groups namely green, brown and red algae based on the pigmentation, morphological and anatomical characters. Seaweeds are one of the commercially important marine living and renewable resources of our country. They contain more than 60 trace elements, minerals, protein, iodine, bromine, vitamins and several bioactive substances hence, are of great economic value and also they serve as both feeding and breeding grounds for invertebrates and fishes. There are 20,000 species are recorded in world, In India 833 species have been recorded.

The total standing crop of seaweeds in the intertidal and shallow waters is 91,339 tones (Wet wt.) consisting of 6,000 tones of agar yielding seaweeds, 16,000 tones of algin yielding seaweeds and remaining edible and other seaweeds (Kaliaperumal, 2000).

Seaweeds are used as food, animal fodder, meal and manure. Antitumor activity, antimicrobial activity, anti-hypercholesterolemic activity, anti coagulant substance, immunomodulating activity, immunosuppressive activity and anti-ulcer activity are mentioned by (Anantharaman *et al*, 2006).

Uses of Seaweeds: An Overview

The use of seaweed as food has been traced back to the fourth century in Japan and the sixth century in China. Today those two countries and the Republic of Korea are the largest consumers of seaweed as food. However, as nationals from these countries have migrated to other parts of the world, the demand for seaweed for food has followed them, as, for example, in some parts of the United States of America and South America. Increasing demand over the last fifty years outstripped the ability to supply requirements from natural (wild) stocks. Research into the life cycles of these seaweeds has led to the development of cultivation industries that now produce more than 90 per cent of the market's demand. In Ireland, Iceland and Nova Scotia (Canada), a different type of seaweed has traditionally been eaten, and this market is being developed. Some government and commercial organizations in France have been promoting seaweeds for restaurant and domestic use, with some success. An informal market exists among coastal dwellers in some developing countries where there has been a tradition of using fresh seaweeds as vegetables and in salads.

China is the largest producer of edible seaweeds, harvesting about 5 million wet tonnes. The greater part of this is for Kombu, produced from hundreds of hectares of the brown seaweed, *Laminaria japonica*, that is grown on suspended ropes in the ocean. The Republic of Korea grows about 800 000 wet tonnes of three different species, and about 50 per cent of this is for *wakame*, produced from a different brown seaweed, *Undaria pinnatifida*, grown in a similar fashion to *Laminaria* in China. Japanese production is around 600 000 wet tonnes and 75 per cent of this is for nori, the thin dark seaweed wrapped around a rice ball in sushi. Nori is produced from red seaweed - a species of *Porphyra*. It is a high value product, about US$ 16 000/dry tonne, compared to kombu at US$ 2 800/dry tonne and wakame at US$ 6 900/dry tonne.

Various red and brown seaweeds are used to produce three hydrocolloids: agar, alginate and carrageenan. A hydrocolloid is a non-crystalline substance with very large molecules and which dissolves in water to give a thickened (viscous) solution. Alginate, agar and carrageenan are water-soluble carbohydrates that are used to thicken (increase the viscosity of) aqueous solutions, to form gels (jellies) of varying degrees of firmness, to form water-soluble films, and to stabilize some products, such as ice cream (they inhibit the formation of large ice crystals so that the ice cream can retain a smooth texture).

Seaweeds as a source of these hydrocolloids dates back to 1658, when the gelling properties of agar, extracted with hot water from red seaweed, were first discovered in Japan. Extracts of Irish Moss, red seaweed, contain carrageenan and were popular as thickening agents in the nineteenth century. It was not until the 1930s that extracts of brown seaweeds, containing alginate, were produced commercially and sold as thickening and gelling agents. Industrial uses of seaweed extracts expanded rapidly after the Second World War, but were sometimes limited by the availability of raw materials. Once again, research into life cycles has led to the development of cultivation industries that now supply a high proportion of the raw material for some hydrocolloids.

Today, approximately 1 million tonnes of wet seaweed are harvested and extracted to produce the above three hydrocolloids. Total hydrocolloid production is about 55 000 tonnes, with a value of US$ 585 million.

Alginate production (US$ 213 million) is by extraction from brown seaweeds, all of which are harvested from the wild; cultivation of brown seaweeds is too expensive to provide raw material for industrial uses. Agar production (US$ 132 million) is principally from two types of red seaweed, one of which has been cultivated since the 1960-70s, but on a much larger scale since 1990, and this has allowed the expansion of the agar industry.

Industrial Uses of Seaweeds

Agar-Agar

Agar is the major constituent of the cell walls of certain red algae, especially members of the *Gelidiaceae* and *Gracilariaceae*. Agar is the Malay world for the gelling substance extracted from *Eucheuma,* but now know to be a carrageenan. The term agar is now generally applied to those algal galactans which have agarose, the disaccharide agarobiose, as their repeating unit. A family of molecules varying amounts of agarose and agaropectin forming the polymers has been described by various investigators (Araki 1978; Duckworth *et al.*, 1971; Izumi 1971 and 1973; Rees 1972; Craigie and Leigh 1978). Agar is soluble in and extracted by hot water; it gels at room temperature and freezing and thawing are used to purify it. The best known use of agar is as a solidifying agent in culture media in bacteriology.

Two genera, *Gelidium* and *Gracilaria*, account for most of the raw material used for the extraction of agar. Extraction of *Gelidium* species gives the higher quality agar (as measured by the gel strength: the strength of a jelly formed by a 1.5 per cent solution).

All *Gelidium* used for commercial agar extraction comes from natural resources, principally from France, Indonesia, the Republic of Korea, Mexico, Morocco, Portugal and Spain. *Gelidium* is a small, slow growing plant and while efforts to cultivate it in tanks/ponds have been biologically successful, it has generally proved to be uneconomic. However, one company, Marine Byproducts International, has launched high-grade agar and agarose products that they claim are derived from their own cultivated *Gelidium*. Presumably the profit from these products at the high end of the market is sufficient to offset the costs of cultivation. Small quantities of *Gracilariopsis* are harvested in Chile and species of *Gelidiella* provide the raw material for a small agar industry in India.

Gracilaria species were once considered unsuitable for agar production because the quality of the agar was poor (gel strength too low). In the 1950s, it was found that pre-treatment of the seaweed with alkali before extraction lowered the yield but gave a good quality agar. This allowed expansion of the agar industry, previously limited by the supply of *Gelidium* available, and led to the harvesting of a variety of wild species of *Gracilaria* in countries such as Argentina, Chile, Indonesia and Namibia. Chilean *Gracilaria* was especially useful, but soon there was evidence of over harvesting of the wild crop. Cultivation methods were then developed, both in ponds and in the open waters of protected bays. These methods have spread beyond Chile to other countries, such as China, the Republic of Korea, Indonesia, Namibia, the Philippines and Viet Nam, usually using species of *Gracilaria* native to each particular country. Obviously, *Gracilaria* species can be grown in both cold and warm waters. Today the supply of *Gracilaria* still comes mainly from the wild, with the degree of cultivation depending on price fluctuations. The important and commonly occurring agarophytes of India are *Gelidiella acerosa, Gracilaria edulis, G.crassa, G. corticata* and *G. folifera*.

- ☆ The stronger and resistant gels are produced at lower concentrations (1 to 1.5 per cent) with water. There is no need to add reagents to produce gelation, such as potassium (protein as necessary with carrageenan) calcium (divalent cations as necessary with alginates).
- ☆ It can be used over a wide range of pH from 5 to 8 and in some cases beyond these limits
- ☆ It withstands thermal treatments very well, even above 100°C which allows good sterilization
- ☆ It assimilates and enhances flavours of products mixed with it and act as a fragrance-fixer permitting their long term fixation
- ☆ Its gel has an excellent reversibility allowing it to be repeatedly gelled and melted without losing any of the original properties

- ☆ Transparent gels that are easily coloured can be obtained whose refractive index can also be easily increased by adding sugar, glucose, glycerine etc., given them an attractive brightness
- ☆ The gel is very stable, not causing precipitates in the presences of certain cations as happens to alginates with calcium.

Agar-Agar Production Method (Flow Chart I to III)

A short and simplified description of the extraction of agar from seaweeds is that the seaweed is washed to remove foreign matter and then heated with water for several hours. The agar dissolves in the water and the mixture is filtered to remove the residual seaweed. The hot filtrate is cooled and forms a gel (jelly) which contains about 1 per cent agar. The gel is broken into pieces, and sometimes washed to remove soluble salts, and, if necessary, it can be treated with bleach to reduce the colour. Then the water is removed from the gel, either by a freeze-thaw process or by squeezing it out using pressure. After this treatment, the remaining water is removed by drying in a hot-air oven. The product is then milled to a suitable and uniform particle size.

There are some differences in the treatment of the seaweed prior to extraction, depending on the genus used. *Gelidium* is simply washed to remove sand, salts, shells and other foreign matter and is then placed in tanks for extraction with hot water. *Gracilaria* is also washed, but it must be treated with alkali before extraction; this alkaline pre-treatment causes a chemical change in the agar from *Gracilaria*, resulting in an agar with increased gel strength. Without this alkaline pre-treatment, most *Gracilaria* species yield an agar with a gel strength that is too low for commercial use. For the alkali treatment, the seaweed is heated in 2-5 per cent sodium hydroxide at 85-90°C for 1 hour; the strength of the alkali varies with the species and is determined by testing on a small scale. After removal of the alkali, the seaweed is washed with water, and sometimes with very weak acid to neutralize any residual alkali.

For the hot-water extraction, *Gelidium* is more resistant and extraction under pressure (105-110°C for 2-4 hours) is faster and gives higher yields. *Gracilaria* is usually treated with water at 95-100°C for 2-4 hours. The remainder of the process is the same for both types of raw material. The hot extract is given a coarse filtration to remove the seaweed residue, filter aid is added and the extract is pumped through a filter press equipped with a fine filter cloth. The extract is thick and will gel if allowed to cool, so it must be kept hot during the filtration processes.

The filtrate is now cooled to form a gel, which is broken into pieces. This gel contains about 1 perc ent agar. The remaining 99 per cent is water that may contain salts, colouring matter and soluble carbohydrates. The gel may be treated with bleach to reduce any colour, washed to remove the bleach, and allowed to soak in water so that most of the salts can be removed by osmosis. The wash waters are drained and the remainder of the process is concerned with the removal of the 99 per cent water in the gel. Either of two methods can be used for this.

The original method of water removal is the freeze-thaw process. The gel is slowly frozen so that large ice crystals form. The structure of the gel is broken down by the freezing so that when the material is thawed most of the water drains away, leaving a concentrated gel that now contains about 10-12 per cent agar (this means about 90 per cent of the original water content has been removed, and with it went a high proportion of any salts, soluble carbohydrates and soluble proteins that may have been present in the gel). Sometimes this gel is placed between porous filter cloths and squeezed in a hydraulic press to remove more water. However, this is a slow process, and usually the thawed material is simply drained and placed in a hot-air dryer. After drying it is milled to the required particle size, usually about 80-100 mesh size. Because of the refrigeration costs, this freeze-thaw process is relatively expensive, compared to the alternative described next.

Flow Chart I: Production of Agar on Commercial Scale in India (Kaliaperumal and Uthirasivam, 2001)

Dried Gracilaria edulis (200 kg) leaching in freshwater in for 12-18 hrs

↓

Washing 2 times in freshwater in agitator tanks (3 washes–10 minutes duration each)

↓

Softening of seaweeds with Hcl for 12 hours (pH 2-4)

↓

Washing with freshwater (pH 7)

↓

Cooking seaweed in digester for 2-3 hrs by passing steam at 50lb pressure

↓

Settling agar gel for ½ hr

↓

Filtering agar gel through filter cloth

↓

Collection of agar gel in aluminium trays

↓

Cooling agar gel in aluminium trays

↓

Cooling agar gel at room temperature for 1-2 hrs

↓

Shredding agar gel with gel chopper

↓

Freezing agar gel for 24 hrs (till temperature comes down to 20°C)

↓

Thawing agar after removal from freezing room

↓

Drying agar in sun on velon screen frames

↓

Bleaching agar in 10 per cent chlorine water for 5-10 minutes

↓

Washing agar with freshwater for 2-3 times

↓

Drying agar in sun on velon screen frames

↓

Dried agar (15 kg)

↓

Packaging dried agar sheets in polythene bags (1 kg packets)

Flow Chart II: Method for Agar Manufacture on Commercial Scale in India (Visweswara *et al.*, 1965)

Raw material (8.5 kg pulverized seaweeds)

↓

Washing in freshwater

↓

Soaking overnight in freshwater

↓

Wet grinding for 30 minutes in edge runner or pestle-mortar type grinder
Washing with soft water

↓

Seaweed pulp

↓

Extraction of 100 litres of water for 2 hrs in a double jacked open pan evaporator circulated with steam: adjusting the pH to 6 with 400-500 ml of N Sulphuric acid

↓

Filtering (Double jacked vacuum filter)

↓

Cooling the filtrate at room temperature

↓

Shredding the agar gel in gel chopper

↓

Freezing agar gel for 24 hrs (in ice plant)

↓

Thawing (at room temperature)

↓

Drying (in air or in an oven with hot air circulation)

↓

Agar-agar (1 kg yield)

Algin or alginic acid is a major constituent of all brown algae. It is a polymer of D-mannuronic acid and L-guluronic acids; the various salts of alginic acid are termed 'alginates'. Alginic acid and its salts with divalent and trivalent metal irons are generally insoluble in water, while alkali metal salts are water soluble. Algin products are used as binders, stabilizers, emulsifiers, and moulding materials in the pharmaceutical industry, in cosmetics and sops, in dental and food technology, in bakery and candy products, in dairy products, and in fish, meat, sausage and beverage processing. They are also used in a wide range of industrial products including dyes, paints and other coatings, in binding briquettes and explosives, in producing paper and cardboard products, in filters and absorbents, in textile production, in pesticides, polishes and lubricants, in fire retardants and extinguishers, in enameling and ceramics and in other miscellaneous applications(Levring *et al., 1969*).

Flow Chart III: Agar Production in Small Scale in India (Mathew, 1999)

Seaweeds
↓
Washing with water
↓
Treating with alkali
↓
Washing of seaweeds
↓
Extracting agar from seaweeds by boiling with under pressure
↓
Filtration
↓
Gel formation
↓
Scoring of gel
↓
Freezing and thawing of synersis
↓
Bleaching with chlorine water
↓
Washing
↓
Drying and powdering
↓
Algin or Alginic acid

Algin Production Method (Flow Chart IV to VI)

The first is to add acid, which causes alginic acid to form; this does not dissolve in water and the solid alginic acid is separated from the water. The alginic acid separates as a soft gel and some of the water must be removed from this. After this has been done, alcohol is added to the alginic acid, followed by sodium carbonate which converts the alginic acid into sodium alginate. The sodium alginate does not dissolve in the mixture of alcohol and water, so it can be separated from the mixture, dried and milled to an appropriate particle size that depends on its particular application.

The second way of recovering the sodium alginate from the initial extraction solution is to add a calcium salt. This causes calcium alginate to form with a fibrous texture; it does not dissolve in water and can be separated from it. The separated calcium alginate is suspended in water and acid is added to convert it into alginic acid. This fibrous alginic acid is easily separated, placed in a planetary type mixer with alcohol, and sodium carbonate is gradually added to the paste until all the alginic acid is converted to sodium alginate. The paste of sodium alginate is sometimes extruded into pellets that are then dried and milled.

Flow Chart IV: Processing technology for Alginic acid production (Pillai, 1957)

Dry seaweed

↓ ← Washing with water

Boiling with water (30 minutes)

↓

Boiling with 0.3 HCl (at 100° C for 30 minutes)

↓ ← Washing with water

Digestion at room temperature

↓

(With 2 per cent soda ash solution for 1-2 hrs)

↓

Filtration

↓ ← Dilution with water

Crude Sodium Alginate

↓

Precipitation of Alginic Acid (With 5 per cent HCl at pH of 2.5)

↓

Filtration

↓

Bleaching (With 2 per cent Potassium Permanganate and 5 per cent HCl)

↓

Filtration

↓ ← Washing with water

Alginic Acid

The process appears to be straightforward, certainly the chemistry is simple: convert the insoluble alginate salts in the seaweed into soluble sodium alginate; precipitate either alginic acid or calcium alginate from the extract solution of sodium alginate; convert either of these back to sodium alginate, this time in a mixture of alcohol and water, in which the sodium salt does not dissolve. The difficulties lie in handling the materials encountered in the process, and to understand these problems a little more detail of the process is required.

Flow Chart V: Processing Technology for Calcium Alginate Production on Cottage Industry (Sadasivan Pillai, 1961)

Fresh seaweed (10 lbs)

↓ ← Drying and pulverizing the weed with wooden mortar and pestle

Dry seaweed powder (25 lbs)

↓

Pretreatment (in earthern pots for about 12 hrs with 100°C of HCl diluted with 1 gallon of water)

↓ ← Washing with water

Extraction (4 hrs at room temperature with 0.25lb of sodium carbonate dissolved in 1 gallon of water)

↓ ← Dilution with water

Water and seaweed residue → Filtration (with a cloth bag)

↓

Bleaching (with 10 per cent acetic acid solution and 1.5 per cent solution of sodium hypochlorite)

↓

Precipitation (with 1 litre of 10-15 per cent calcium chloride solution)

↓ ← Washing with water

Calcium Alginate

To extract the alginate, the seaweed is broken into pieces and stirred with a hot solution of an alkali, usually sodium carbonate. Over a period of about two hours, the alginate dissolves as sodium alginate to give very thick slurry. This slurry also contains the part of the seaweed that does not dissolve, mainly cellulose. This insoluble residue must be removed from the solution. The solution is too thick (viscous) to be filtered and must be diluted with a very large quantity of water. After dilution, the solution is forced through a filter cloth in a filter press. However, the pieces of undissolved residue are very fine and can quickly clog the filter cloth. Therefore, before filtration is started, a filter aid, such as diatomaceous earth, must be added; this holds most of the fine particles away from the surface of the filter cloth and facilitates filtration. However, filter aid is expensive and can make a significant contribution to costs. To reduce the quantity of filter aid needed, some processors force air into the extract as it is being diluted with water (the extract and diluting water are mixed in an in-line mixer into which air is forced). Fine air bubbles attach themselves to the particles of residue. The diluted extract is left standing for several hours while the air rises to the top, taking the residue particles with it. This frothy mix of air and residue is removed from the top and the solution is withdrawn from the bottom and pumped to the filter. The next step is precipitation of the alginate from the filtered solution, either as alginic acid or calcium alginate.

Flow Chart VI: Method for Extraction of Sodium Alginate, Calcium Alginate and Alginic Acid (Visweswara Rao and Mody, 1964)

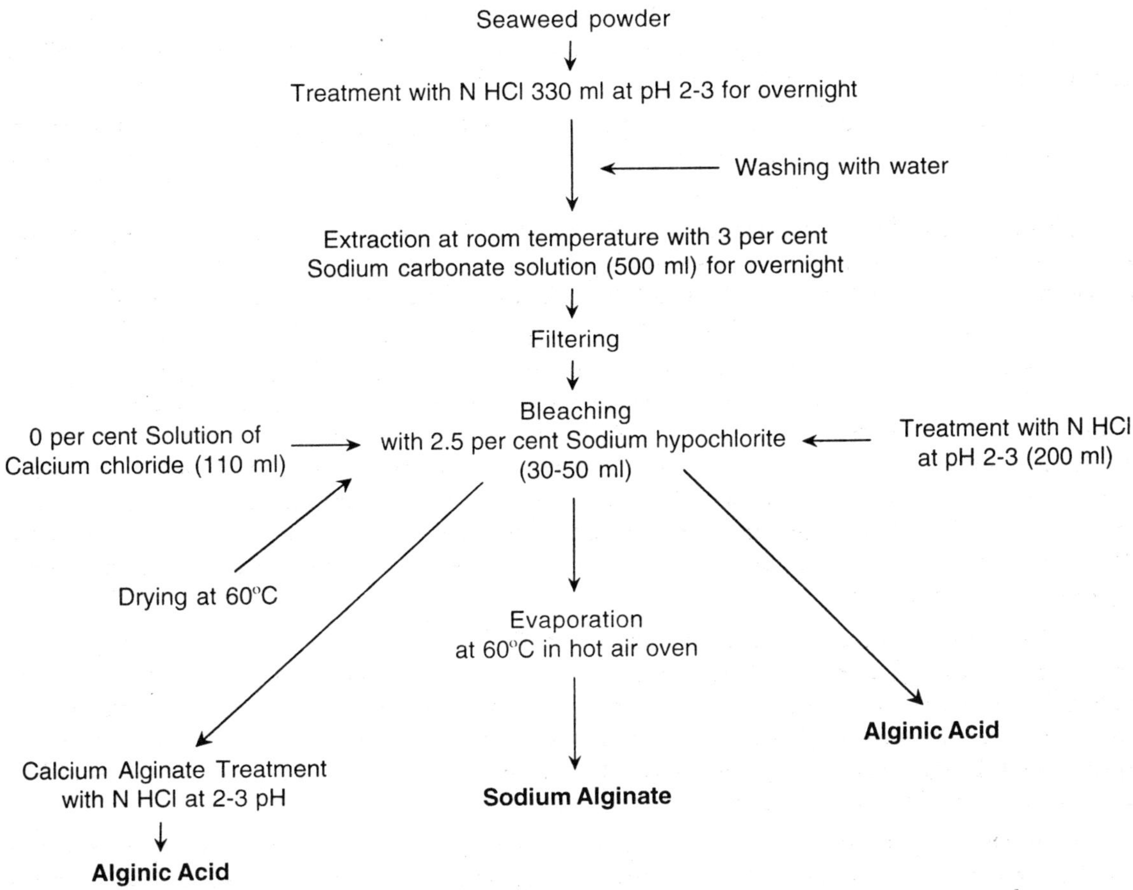

Carrageenan

Carrageenan is sulphated galactan polymer obtained from various red seaweeds belonging to the families Gigartinaceae, Solieriaceae and Hypneaceae.It differs from agar mainly in its higher sulphated fraction and higher ash content. The back bone of the carrageenan polymer consists of 1,3 and 1,4 linked D-galactopyaranose units which vary in the degree and the location of sulphated esterification. Carrageenan can be separated in to two fractions-k-carrageenan and l-carrageenan whose polymer chain is branched in the former and linear in the latter. K-Fraction is separated from l-fraction by precipitation with potassium chloride and amounts to 40 per cent of the carrageenan, the balance being the l-fraction. The fraction soluble in hot water stands for K- carrageenan and the cold water soluble fraction top are l-carrageenan.

Carrageenan is usually employed as either the sodium, potassium or calcium salt. Pottassium carrageenan as prepared commercially is a mixture of l and k-carrageenan which is soluble in hot water but only the k-carrageenan is gel forming. Sodium carrageenan is soluble in cold water and dose not gel. Because of its gelling properties-carrageenan is more effective as a stabilizer than - l-carrageenan.The properties of these two compounds present in extractives from the various algal species is, therefore, highly important in determining their uses.

In general, when carrageenan is added to flour, starches or albumins there is an increase in their gel strength and viscosity. Addition to gums and alginates gives a decrease in viscosity, whilst if added to agar it results in lowered gel strength.

Blending with other compounds overcome the cohesive, brittle characteristics of the substances. A very useful blend consists of 50 per cent carrageenan, 331/3 per cent locust bean gum and 16 2/3 per cent of a potassium salt, usually KCl.

Under usual conditions the extractives will withstand high heat ad low temperatures provided the humidity is kept low and the pH is at 7.0 or higher. Once the compounds have been put completely into solution they will withstand violent agitation. An exception to this is the chocolate milk system. Oxidizing and reducing agents are destructive.

Carrageenan Inc. of the U.S.A market a whole range of extracts with gel strength ranging from slight to 900g(i-5 per cent in water, measured at 25 per cent with a 0.73 in cylindrical plunger).With the addition of potassium salts the gel strength is further increased. The refined extractives are used in a variety of preparations. They include hand lotions, mineral emulsions, chocolate milks, cream stabilizers, toothpaste, cough syrup, milk base puddings, ice cream stabilizers, etc. The extractive plus potassium salt is used for ice-cream syrups and for tablet binding in pharmaceuticals. When locust beam gum is added the product is used for glazes on baked goods thickening of fruit pie fillings, jellies, preserves, aspics and so on.

Bleached moss is principally used in the preparation of blancmanges and moulds and is often seen in 'health stores' as part of dietetic products. One way of using it is to soak it in water for 10 min, and then let it simmer for 15 min in some milk with a flavoring added to taste. It of finally strained into a basin and after adding some honey or sugar allowed setting. The addition of cream is said to be an improvement. If water, with the addition of fruit juice, is employed instead of milk, a nice fruit jelly is the result. Because of these jellying properties.

As a means of suspending fine cocoa powder in milk carrageenan has to a large extent replaced algin. It has been used for feasting (one glass of jelly in milk per feed) with great success and improved the coats of red setter dogs.

In the textile industry carrageenan is extensively used at a concentration of about 5 per cent as a stiffening and binding material. I produces a sift finish and as surface to which printing will adhere. It is also used to stiffen and provide a gloss printing to leather goods. Leather manufactures require a certain amount of carrageenan annually for smoothing leather and giving it a gloss and stiffines. The gelose is melted and brushed on to the leather, which is then polished with glass cylinders. Dried plants are also used extensively in shoe polishes, because the mucilage holds down and smooths out the tiny rough projections on the surface of the shoe leather. In Great Britain 100 tonnes or more are used in the manufacture of cold water or casein paints in order to hold the film on the surface of casein dries out. It can also be used to bind briquettes of vegetable charcoal powder.

Apart from toothpastes and hand lotions it is also used in the production of shaving soaps and hair creams. For trade purposes 'carrageenan naturally' 'carrageenan depuratum 'and 'carrageenan electrum albissimum'. The last named is the best-known pharmaceutical emulsifier, Small quantities of benzoic acid or sodium benzoate are added to this a preservative. It is important from the medical point of view, is that carrageenan, even in very great dilution, acts as an anti-coagulant for blood, Because of its mucus forming properties it was used in diseases of the lungs and also to correct the taste of bitter drugs. A preparation made from this alga was given to soliers who had been gassed in the First World War, presumably because it eased the throat. Liver oil emulsions of carrageenan have been made whilst interesting preparations is cotton-wool soaked in a carrageenan decoction and dried. Same time the culture is best place for hiding, breeding area fishes, food for herbivore fishes and associated organisms.

Carrageenan production (US$ 240 million) was originally dependent on wild seaweeds, especially Irish Moss, a small seaweed growing in cold waters, with a limited resource base. However, since the early 1970s the industry has expanded rapidly because of the availability of other carrageenan-containing seaweeds that have been successfully cultivated in warm-water countries with low labour costs. Today, most of the seaweed used for carrageenan production comes from cultivation, although there is still some demand for Irish Moss and some other wild species from South America.

Seaweed meal, used an additive to animal feed, has been produced in Norway, where its production was pioneered in the 1960s. It is made from brown seaweeds that are collected, dried and milled. Drying is usually by oil-fired furnaces, so costs are affected by crude oil prices. Approximately 50 000 tonnes of wet seaweed are harvested annually to yield 10 000 tonnes of seaweed meal, which is sold for US$ 5 million.

Fertilizer uses of seaweed date back at least to the nineteenth century. Early usage was by coastal dwellers, who collected storm-cast seaweed, usually large brown seaweeds, and dug it into local soils. The high fibre content of the seaweed acts as a soil conditioner and assists moisture retention, while the mineral content is a useful fertilizer and source of trace elements. In the early twentieth century, a small industry developed based on the drying and milling of mainly storm-cast material, but it dwindled with the advent of synthetic chemical fertilizers. Today, with the rising popularity of organic farming, there has been some revival of the industry, but not yet on a large scale; the combined costs of drying and transportation have confined usage to sunnier climates where the buyers are not too distant from the coast.

The growth area in seaweed fertilizers is in the production of liquid seaweed extracts. These can be produced in concentrated form for dilution by the user. Several can be applied directly onto plants or they can watered in, around the root areas. There have been several scientific studies that prove these products can be effective. In 1991, it was estimated that about 10 000 tonnes of wet seaweed were used to make 1000 tonnes of seaweed extracts with a value of US$ 5 million. However, the market has probably doubled in the last decade because of the wider recognition of the usefulness of the products and the increasing popularity of organic farming, where they are especially effective in the growing of vegetables and some fruits.

Cosmetic products, such as creams and lotions, sometimes show on their labels that the contents include "marine extract", "extract of alga", "seaweed extract" or similar. Usually this means that one of the hydrocolloids extracted from seaweed has been added. Alginate or carrageenan could improve the skin moisture retention properties of the product. Pastes of seaweed, made by cold grinding or freeze crushing, are used in thalassotherapy, where they are applied to the person's body and then warmed under infrared radiation. This treatment, in conjunction with seawater hydrotherapy, is said to provide relief for rheumatism and osteoporosis.

Over the last twenty years there have been some large projects that investigated the possible use of seaweeds as an indirect source of fuel. The idea was to grow large quantities of seaweed in the ocean and then ferment this biomass to generate methane gas for use as a fuel. The results showed the need for more research and development, that it is a long-term project and is not economic at present.

There are potential uses for seaweed in wastewater treatment. Some seaweed is able to absorb heavy metal ions such as zinc and cadmium from polluted water. The effluent water from fish farms usually contains high levels of waste that can cause problems to other aquatic life in adjacent waters. Seaweeds can often use much of this waste material as nutrient, so trials have been undertaken to farm seaweed in areas adjacent to fish farms.

Carrageenan Production Methods (Flow Chart VII to XI)

There are two different methods of producing carrageenan, based on different principles.

In the original method, the only one used until the late 1970s-early 1980s - the carrageenan is extracted from the seaweed into an aqueous solution, the seaweed residue is removed by filtration and then the carrageenan is recovered from the solution, eventually as a dry solid containing little else than carrageenan. This recovery process is difficult and expensive relative to the costs of the second method.

In the second method, the carrageenan is never actually extracted from the seaweed. Rather the principle is to wash everything out of the seaweed that will dissolve in alkali and water, leaving the carrageenan and other insoluble matter behind. This insoluble residue, consisting largely of carrageenan and cellulose, is then dried and sold as semi-refined carrageenan (SRC). Because the carrageenan does not need to be recovered from solution, the process is much shorter and cheaper.

Refined Carrageenan and Filtered Carrageenan

Refined carrageenan is the original carrageenan and until the late 1970s-early 1980s was simply called carrageenan. It is now sometimes called filtered carrageenan. It was first made from *Chondrus crispus*, but now the process is applied to all of the above algae.

The seaweed is washed to remove sand, salts and other foreign matter. It is then heated with water containing an alkali, such as sodium hydroxide, for several hours, with the time depending on the seaweeds being extracted and determined by prior small-scale trials, or experience. Alkali is used because it causes a chemical change that leads to increased gel strength in the final product. In chemical terms, it removes some of the sulphate groups from the molecules and increases the formation of 3,6-AG: the more of the latter, the better the gel strength. The seaweed that does not dissolve is removed by centrifugation or a coarse filtration, or a combination. The solution is then filtered again, in a pressure filter using a filter aid that helps to prevent the filter cloth becoming blocked by fine, gelatinous particles. At this stage, the solution contains 1-2 per cent carrageenan and this is usually concentrated to 2-3 per cent by vacuum distillation and ultrafiltration.

Semi-Refined Carrageenan and Seaweed Flour

Semi-refined carrageenan (SRC) was the name given to the product first produced by the second method of processing noted in next paragraphs. This is the method in which the carrageenan is never actually extracted from the seaweed.

Flow Chart VII: Processing Technology for Kappa Carrageenan Production (Ji Ming Hou, 1990)

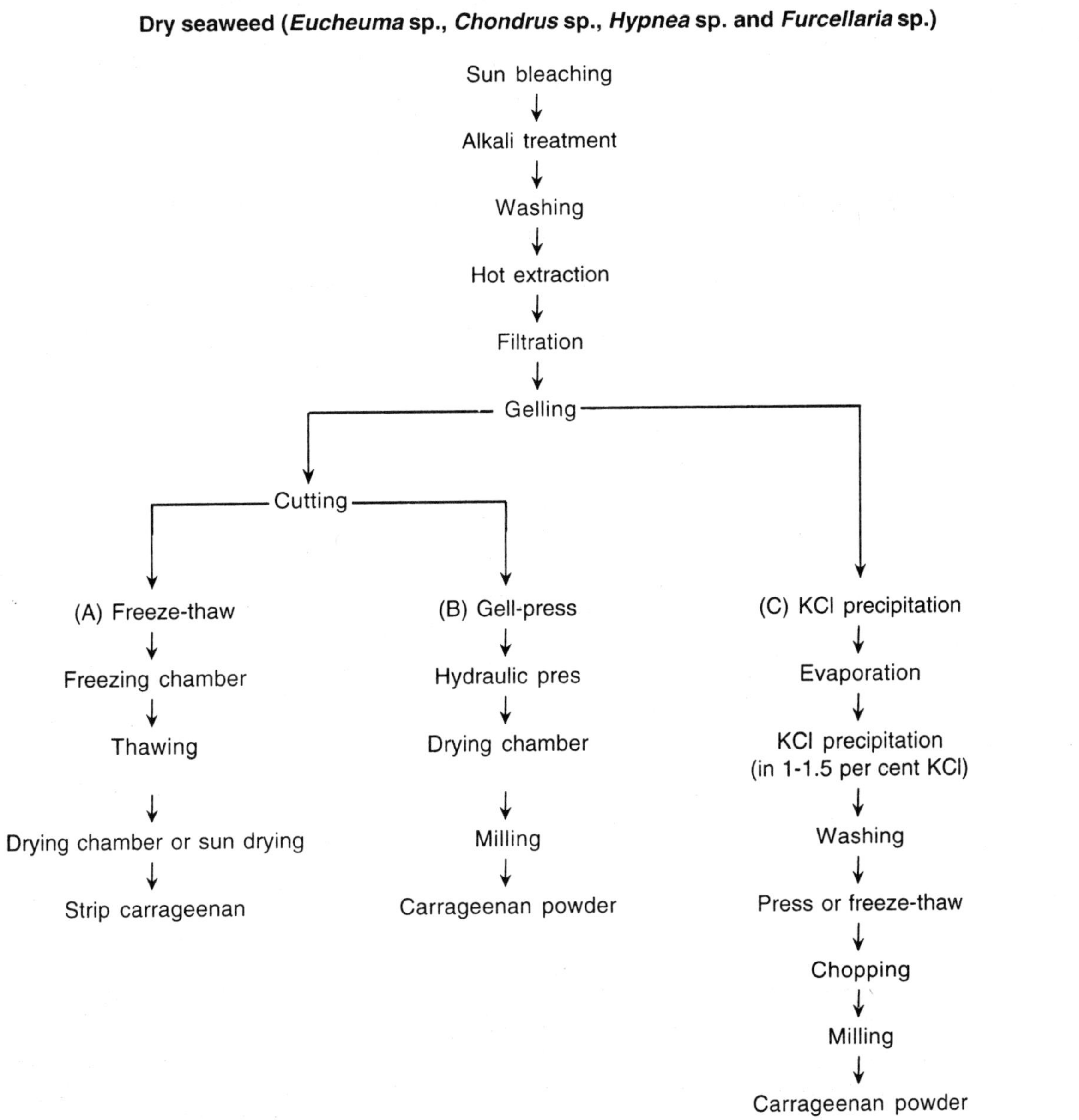

In the production of SRC, *Kappaphycus alvarezii*, contained in a metal basket, is heated in an alkaline solution of potassium hydroxide for about two hours. The hydroxide part of the reagent penetrates the seaweed and reduces the amount of sulphate in the carrageenan, increases the 3,6-AG so the gel strength of the carrageenan in the seaweed is improved. The potassium part of the reagent combines with the carrageenan in the seaweed to produce a gel and this prevents the carrageenan from dissolving in the hot solution. However, any soluble protein, carbohydrate and salts do dissolve and are removed when the solution is drained away from the seaweed. The residue, which still looks like seaweed, is washed several times to remove the alkali and anything else that will dissolve in the water. The alkali-treated seaweed is now laid out to dry; in hot climates, like the Philippines, usually on a large concrete slab. After about two days it is chopped and fed into a mill for grinding to the powder that is sold as SRC or seaweed flour.

Flow Chart VIII: Processing Technology for Lota and Lamda Carrageenan Production (Ji Ming Hou, 1990)

Dried seaweed (*Gigartina* sp., and *Eucheuma* sp.)

Sun bleaching
↓
Washing
↓
Hot extraction
↓
Filtration
↓
Evaporation
↓
Alcohol precipitation
↓
Centrifuge
↓
Drying
↓
Milling
↓
Lota or Lamda carrageenan powder

However, the seaweed flour is coloured, often has a high bacterial count and is not suitable for human consumption. Nevertheless it immediately found a large market in canned pet food because it is a good gelling agent and was so much cheaper than refined carrageenan. The temperatures used in the canning process destroy any bacteria so the high bacterial count in the SRC is not a problem. Sometimes the dried product is just chopped into pieces, not milled, and sold as a raw material to refined carrageenan processors. It is called alkali treated cottonii (ATC) or alkali treated cottonii chips (ATCC), or even simply cottonii chips. If this treatment is done in the country of origin of the seaweed, such as the Philippines or Indonesia, this means processors in Europe and United States of America have cheaper transport costs per tonne of carrageenan, compared with shipping dried seaweed. They have also left behind some waste products, which reduce their waste treatment costs.

Flow Chart IX: Processing Technology for Semi Processed-Carrageenan (Ji Ming Hou, 1990)

Dried Seaweed
(*Kappaphycus alvarezii, Eucheuma cottonii*)

↓

Cutting

↓

Alkali treatment
(8.5 per cent KOH)

↓

Rinsing

↓

Sun drying

↓

Milling

↓

Semi-processed carrageenan product

Flow Chart X: Processing Technology for Crude-Carrageenan (Gopakumar, 1997)

Washed and sun dried seaweed
(*Eucheuma* sp.)

↓

Digestion with 0.5 per cent KOH at 90°C for 60 minutes

↓

Alkali digested seaweed

↓

Washing with clean water

↓

Alkali free digested seaweed

↓

Sun drying

↓

Pulverization in grinding mill/Milling

↓

Powdered seaweed (Yellow colour)

↓

Crude carrageenan

Flow Chart XI: Processing Technology for Semi-Refined Carrageenan (Kaliaperumal, 1984)

Cleaned dried *Eucheuma*

↓

Washing the cleaned Seaweed

↓

In fresh water

↓

Boil the seaweed

↓

Treat with 10 per cent Potassium hydroxide

↓

(Caustic potash)

↓

Sun drying

↓

Grinding

↓

Powdered semi-refined carrageenan

Kappaphycus alvarezii is used in this process because it contains mainly kappa carrageenan and this is the carrageenan that forms a gel with potassium salts. Iota-containing seaweeds can also be processed by his method, although the markets for iota carrageenan are significantly less than those for kappa. Lambda carrageenan do not form gels with potassium and would therefore dissolve and be lost during the alkali treatment.

The simplicity of the process means the product is considerably cheaper than refined carrageenan.

There is no alcohol involved that must be recovered, no distillation equipment to purify alcohol, no equipment for making gels, no refrigeration to freeze the gels, nor any expensive devices to squeeze the water from the gel.

Medicinal Uses

Seaweeds were considered to be of medicinal value in the orient as early as 3000 B.C. The Chinese and Japanese used them in the treatment of goiter and other glandular diseases. The Romans used the seaweeds for healing the wounds, burns and rashes. The British used the *Porphyra* to prevent scurvy (vitamin deficiency disease) during long voyages.

Various red algae particularly *Coralline officinalis* and *Alsidium helminthocorton* were employed as vermifuges. Some other red algae such as *Chondrus, Gracilaria, Gelidium* and *Pterocladia* have been used to treat various stomach and intestinal disorders and also helped to relieve from constipation and other discomforts. *Laminaria* used as a pain- killer and also used to distend the uterus. A number of species of marine algae have been found to have the anticoagulant and antibiotic properties. Carrageenan may be useful in ulcer therapy and the alginates are found to prolong the rate of activity in certain drugs. Species of *Sargassum* were used for cooling and blood cleaning effect. *Hypnea musciformis used* as a vermifuge or worm expelling agent. The iodine rich seaweed *Sarconema* can be used for controlling the goiter disease caused by the enlargement of thyroid gland.

Though the importance of different seaweed products in pharmacology is known, the development of antibacterial, antifungal and antiviral substances from seaweeds is still in a growing stage of research and development. *Gelidium cartilagineum* have been found to be active against influenza B and mumps virus. Among the seaweeds, red algae have been the major producer of bioactive secondary metabolites. Isolation of polysaccharides and other compounds with antiviral activity against enveloped viruses increased interest in algae as a source of antiviral compounds. The antiviral effects of polysaccharides from marine algae towards mumps virus and influenza B virus were reported. Subsequently, polysaccharides from extracts of red algae were found to inhibit herpes simplex virus (HSV) and other viruses. Extracts from the California red algae *Schizymenia pacifica* contained a sulfated polysaccharide in the γ-carrageenan family, which selectively inhibited HIV reverse transcriptase.

Table 14.1: Seaweed Used as a Medicine
(Data from Newton 1951; Boney 1965; Chapman 1970.)

Name	*Use*
Red algae	
Alsidium helminthochorton	Vermifuge
Digenia simplex	Vermifuge
Corallina officinalis	Vermifuge
Chondrus crispus	Coughs, chest, stomach ailments
Brown algae	
Fucus vesculosus	Scrofula
Fucus evanescens	Stomach ailments
Laminaria and other kelps	Iodine source(cures goiter), stipes used to Green algae open wounds, in cervicaldilation
Iva species	Burn treatment
Acetabularia major	Bladder and kidney ailments

Anti-Tumor Activity from Seaweeds

An anti-tumor activity was tested on the different fraction of red alga *Porphyra telfairiae*. It was shown that the petroleum ether fraction showed a obvious inhibition effects on Sarcoma 180 (S sub (180)), Hepatic carcinoma (Hep A) and Ehrlich carcinoma (EAC) under a dose dependent relation in the range of 125-500 mg x kg super (-1) *in vitro*, meanwhile, the inhibitory effects on tumor didn't affect the increasing of the body weight of mice. In addition, the petroleum fraction showed some cell toxicity on HeLa cell and HL sub (60) cell *in vitro*. The chloroform and butanol fraction had a less effects compared with the petroleum ether fraction and the water fraction had a strong toxicity. Five compounds were isolated from petroleum ether fraction, and structures of four were elucidated as β-Sitosterol (I), Stearilacid (II), Tetratriacontyl stearate (III) and β-Sitosterol palmitate (IV) on the basis of chemical and spectral evidence.

Sulfated Polysaccharides from Seaweeds

Fucans, a family of sulfated polysaccharides present in brown seaweed, have several biological activities. Their use as drugs would offer the advantage of no potential risk of contamination with viruses or particles such as prions. A fucan prepared from *spatoglossum schroèderi* was tested as a possible inhibitor of cell-matrix interactions using wild-type chinese hamster ovary cells (cho-k1) and the mutant type deficient in xylosyltransferase (cho-745). The effect of this polymer on adhesion properties with specific extracellular matrix components was studied using several matrix proteins as substrates for cell attachment. Treatment with the polymer inhibited the adhesion of fibronectin to both cho-k1 (2 x 10 super (5)) and cho-745 (2 x 10 super (5) and 5 x 10 super (5)) cells. No effect was detected with laminin, using the two cell types. On the other hand, adhesion to vitronectin was inhibited in cho-k1 cells and adhesion to type i collagen was inhibited in cho-745 cells. In spite of this inhibition, the fucan did not affect either cell proliferation or cell cycle.

Antibacterial Activity from Seaweeds

The antibacterial activity of 57 seaweed species collected in the intertidal and reef zones of 19 coastal localities of the Mexico gulf and Caribbean Sea was studied. Crude extracts were obtained using, ethanol, acetone and water as solvents; they were tested against four pure bacterial strains: *staphylococcus aureus, escherichia coli, shigella sonnei* and *streptococcus pyogenes*. The extracts of 31 species showed some degree of antibacterial activity, from them 16 were highly active showing inhibition zones higher than 10 mm; 15 had moderate activity with inhibition zones lower than 10 mm. From the 31 active species 9, were Chlorophyta, 6 Phaeophyta and 16 Rhodophyta. The variation of antibacterial activity related to the locality or time of collection was observed in: digenea simplex, *Gracilaria cornea, Laurencia intricata, L. Obtusa* and *L. Papillosa*. The extracts of *Liagora farinosa, Dasycladus vermicularis* and *Lobophora variegata* presented the highest inhibition zones and a wide spectrum of antibacterial activity. The solvent polarity makes easy the extraction of antibacterial substances. The acetone was recommended as a suitable solvent for this kind of extractions.

Anti-hypercholesterolemic Activity from Seaweeds

Each right and left sides of the peritoneal cavity, levels of total cholesterol and low-density lipoprotein were decreased to 37 per cent and 24 per cent compared to triton injection only. For histochemical changes, hepatic tissues obtained at 40 hours after injection of the triton and the *Porphyra yezoensis* extract was fixed in fromol-calcium solution. Numbers of lipid drops and cholesterol particles decreased in the portal space of the hepatic cytoplasm. This indicates that the accumulation of lipid, including cholesterol, caused by triton was prevented by the antihypercholesterolemic effect of extract from the seaweed *P. yezoensis*.

Anti-Coagulant Substance from Seaweeds

An anti-coagulant isolated from the marine green alga *Codium pugniformis* was composed mainly of glucose with minor amounts of arabinose and galactose. It was highly sulfated (326 µg mg super (-1) polysaccharide) and contained protein (52 µg mg super (-1) polysaccharide) and was thus a proteoglycan. The anticoagulant properties of the purified proteoglycan were compared with those of heparin by studying the activated partial thromboplastin time (APTT), prothrombin time (PT) and thrombin time (TT) using normal human plasma. The proteoglycan showed similar activities to heparin, but was weaker than heparin. On the other hand, the proteoglycan did not affect PT even at the concentration at which APTT and TT were prolonged. The anticoagulation mechanism of this proteoglycan was due to the direct inhibition of thrombin and the potentiation of antithrombin III.

Ethanol extracts from a group of 53 marine organisms, including a newly identified species, from Baja California sur (mexico), were evaluated for their antimicrobial and antiparasitic activity. The activity against *Staphylococcus aureus*, *Streptococcus faecalis*, *Bacillus subtilis* (gram+), *Escherichia coli* (gram-) and *Candida albicans* (yeast) was determined by the diffusion agar method. *Aplysina gerardogreeni* (Demospongiae) was found to be the most active sample. In addition, *A. gerardogreeni*, *Pacifigorgia media*, and *Pacifigorgia* sp. possess significant activity against *Mycobacterium tuberculosis* h sub (37) RV and *Pacifigorgia media* and *Geodia* sp. against *Mycobacterium avium*. From this group, 15 ethanol extracts were tested in vitro against *Entamoeba histolytica* and *Giardia lamblia*. *Litotamnium crassiussculum*, *Geodia* sp., *Pacifigorgia* sp. showed significant activity against *Entamoeba histolytica* while *Myxilla incrustans* and *Muricea appressa* were active against *Giardia lamblia*. *Litotamnium crassiussculum* showed activity against both trophozoites.

Anti-viral Activity from Seaweeds

Sixteen extracts of British Columbian seaweeds, previously shown to have antiviral activity, were investigated in more detail. In all cases except one (*Analipus japonicus*) the antiviral activity was predominantly virucidal (direct virus-killing effect). The *A. japonicus* extract showed replication-inhibition as well as virucidal activity. All extracts, with the exception of *A. japonicus* again, required light for maximum activity, indicating the presence of antiviral photosensitizers. In some cases light was absolutely required, whereas in others light enhanced the antiviral activity, and furthermore different extracts showed a preference for either visible light or long-wave ultraviolet (UVA). These results taken together indicate the presence of different antiviral compounds with distinct modes of action, and argue strongly against the hypothesis that antiviral activities in seaweeds are due to common ingredients such as polysaccharides. This was further substantiated by the finding that in most cases the antiviral activity could be adsorbed by polyvinylpolypyrrolidone, which indicates the presence of phenolic compounds. In view of the presence of many different antiviral compounds among these extracts, we believe BC seaweeds may provide a source of potentially useful antivirals, and are worthy of further study.

Immunomodulating Activity from Seaweeds

Effect of eight kinds of seaweed extract (SWE) on human lymphocytes was studied in vitro. The extracts of *Hizikia fusiformis* and *Meristotheca papulosa* (green) markedly stimulated human lymphocytes to proliferate, whereas *Eucheuma muricatum* and *Meristotheca papulosa* (red) weakly stimulated proliferation. The responder cells are T cells, because T cells purified by sheep red blood cell (SRBC) rosette-formation were significantly stimulated with SWE, but β cells were not. These extracts enhanced the induction of cytotoxic T lymphocyte (CTL) activity, but failed to enhance natural killer (NK) cell activity. These extracts had a stimulatory effect on immunoglobulin (Ig) production by β cells and tumor necrosis factor (TNF) production by monocytes. The activity of *Hizikia fusiformis* associated with polysaccharides which were extracted with ethanol and purified by ion exchange and gel filtration chromatography, whose molecular weight was about 100 kDa. These results suggest that SWE has an immunomodulating activity on human lymphocytes and this ability might be useful for clinical application to treat several diseases such as tumors.

Biological Activity from Seaweed Extract

The results of the first screening of 89 seaweeds collected from British Columbia, Canada and Korea for antiviral activity. Various concentrations of methanol extracts of dried algae were tested against 100 plaque-forming units of herpes simplex virus type 1 and Sindbis virus in Vero cell monolayers. Eleven extracts inhibited both viruses, and 22 extracts were active against only one of the viruses. Thus, in total 37 per cent of the species were active, and only two of these extracts also showed cytotoxicity at the concentrations tested. The antiviral activities were proportionately more frequent in the Korean extracts (56 per cent compared with 27 per cent of Canadian extracts), but in general the more potent extracts were of Canadian origin. *Analipus japonicus* was the most potent anti-herpes species, and the *Korean* species of *Codium fragile* was the most potent against both viruses. This high yield of antiviral extracts illustrates the potential of seaweeds as a resource for bioactive compounds.

Immunosuppresive Activity from Seaweeds

Water extracts of marine algae with in vitro immunosuppressive activity were investigated for in vivo activity using murine models of collagen-induced arthritis and skin transplantation. Eleven (3 brown and 8 red algae) of them had suppressive activity on the collagen-induced mouse arthritis model. Of these algae, *Eisenia bicyclis, Sargassum sagamianum, Amphiroa aberrans,* and *Gracilaria verrucosa,* in particular, showed high activity. On the other hand, treatment with extracts from *Codium fragile, Codium intricatum, Codium divaricatum,* and *Liagora* sp. prolonged the allograft survival time on the murine skin rejection model. One of these algal extracts, those from Liagora sp., markedly prolonged the allograft survival time. These results suggest that bioactive compounds with immunosuppressive activity may be contained in these algae.

Anti-Ulcer Substance from Seaweeds

The antimicrobial activities of antiulcer substance were studied by the agar streak dilution method using nutrient agar medium for the bacterias and zapek's agar medium for the fungi. The crystal dilutions were made in the acetone and added to the medium in a 1 per cent final concentration. Minimal concentration at which complete inhibition was observed against a series of organisms is shown. In this time which 200 g weighing Wister male rat was used as experimental animals, this crystal substance was shown the bioactivities of prevention against shayulcer. Antiulcer substance from *porphyra tenerakjellman* against stress ulcer.

Other Uses of Seaweeds

Seaweed as a Food for Man

Seaweeds have been used for human food since ancient times. There are historical references to such uses as early as 600–800 B.C. in China, and seaweeds were undoubtedly used in prehistoric times. Seaweeds are eaten for their food value, flavours, and colours, and textures and are typically combined with other type of food. The major edible seaweeds are listed in the table.

Seaweed as a stable item of diet has been used in Japan and China for a very long time. To lesser extent, various species have been employed in Europe and North America, and these areas considered first. On the continents and the larger Islands where normal agriculture can be supported, there is no great demand, except perhaps in less advanced coastal areas, for seaweed as food. There are, however, many Islands where conventional agriculture cannot meet local demands and it is here that people have, of necessity, relied on the sea as a major source of food.

Fresh, dried and processed seaweeds are utilized for human consumption. Many types of seaweed are used as food in Japan, China, Philippines and other countries of Indo-Pacific regions. They are rich in proteins and also contain carbohydrates, vitamins and minerals. Seaweeds are eaten as salad, curry, soup, or jam.

Seaweed as a Animal Fodder

Today, in a number of countries, animals still regularly feed in certain regions upon fresh seaweed or upon a prepared seaweed food. In Iceland fresh seaweed are commonly employed as a food for sheep, cattle and horses; the animals are encouraged to stay browsing on the shore during the summer as well. In many cases the seaweed forms almost their only food, though it is sometimes given along with hay. Horses prefer *Laminaria saccharina,* of which they eat the basal or youngest parts of the frond. Elsewhere in Europe particular seaweed said to be distasteful to cattle. For stall feeding cattle the Icelanders use *Rhodymenia palmata* and *Alaria esculenta,* without the smell or taste of the milk being affected.

In certain coastal area of Norway the sheep are fed regularly on seaweed, and after several germinations it has been found that they digest far better than inland sheep. These accords with experimental evidence on the utilization of seaweed foods *Fucus serratus* and *Chorda filum,* together with the stipes and fronds of *Laminaria digidata,* were given during the winter as on additional food to cattle.

Seaweed Meal

Seaweeds are cheap sources of minerals and trace elements. Hence the meal prepared from seaweeds can be given as supplementary to the daily rations of the cattle, poultry and other farm amimals. Seaweed meals can be obtained by grinding and cleaned and washed seaweeds of *Ulva, Enteromorpha, Sargassum, Padina, Dictyota, Gracilaria* and *Hypnea.* Thivy (1960) described a simple method for the preparation of seaweed meal from *Gracilaria edulis.* Seaweed meal can be mixed with fish meal used as poultry feed. Seaweed have been utilized as animal feed since long back in other countries. But in India scientific approach to this line of study in scarce. Seaweed meal was prepared from *Sargassum* and feeding trials on chicks, sheep and cattle were reported by Dave *et al.,.* (1979). Studies on the feeding *Gracilaria* meal to white leghorn egg laying birds were made by Chaturvedi *et al.* (1979) to find out effect of feeding algae on the quality of egg. They concluded that *Gracilaria* meal at level of 10 per cent can be included in the ration of laying birds replacing yellow maize. Jagannathan Venkata Krishnan (1979) likewise concluded feeding trials replacing ragi *(Eleusine coracana)* with 0, 5, 10 and 15 per cent of seaweed in unsexed day-old white leghorn chicks using six seaweeds commonly available in Tamil Nadu coast.

Seaweed Manure

The value of seaweeds as an agricultural fertilizer has been repeatedly demonstrated, especially by coastal formers with ready access to seaweeds (Booth 1965). While such applications were probabaly practiced in antiquity, the first reference to such use appears in Roman writings from the second century A.D. (Newton 1951). Seaweeds are collected from beach drift or harvested especially for agricultural use.In recent years liquid extracts of brown algae have been appeared on the market. Two well known brands are 'Maxicrop' and 'Alginure'. The former is primarily used for glass-house crops and vegetables whereas the latter is being promoted for pastures and field crops. Maxicrop is obtained by a process of alkaline hydrolysis mainly from fucoides, whereas Alginure is obtained from oar weeds.

Major Genera of Fertilizer Seaweed

Iodine

Lack of iodine can cause developmental structural and neural fetal abnormalities collectively called cretinism. This condition, directly as a result of low maternal iodine supplies, is difficult to correct postpartum, if at all. The treatment is adequate maternal iodine consumption from the mother's initial beginning as an egg in her maternal grandmother. Mammalian fetal iodine need is about three times per unit body weight of the mother. In adult humans, chronic low iodine consumption often results from iodine deficient soils and water, and consequently low iodine food.

The human consequence is: first, goiter, an enlargement of the thyroid gland, deliberately generated by TSH (Thyroid-Stimulating Hormone) to increase thyroid gland cell surface area and more "iodine traps",and secondly, various manifestations of hypothyroidism. The treatment is often simply more dietary iodine for both conditions and this can be easily accomplished with dietary seaweeds. Some seaweeds having iodine such as Icelandic kelp, 8000ppm, Atlantic kelp 1500-2000ppm, Pacific kelps 500-1200ppm, *Fucus* sp. 200-500ppm, *Sargassum* 35ppm, and Nori 15 ppm. These are all approximate and will vary considerably by season, location, age, and harvest practices.

Potassium

Potassium is essential for even minimal nerve and muscle functioning and as a cross-membrane transporter ion for neurotransmittors and hormones. Adding high-potassium foods, especially seaweeds, to the diets of add children (instead of Ritalin) and adults can significantly improve behaviour and mental functioning. Similarly, fibromyalgia patients, the exhausted, the forgetful, the moody, the agitated, anxiety disorders, and depression are all favourably improved with high-potassium diets.

Fucoidan

Fucoidan can be easily cooked out of most edible brown algae by simmering 20-40 minutes in water (alone or in food). When consumed, it seems to reduce the intensity of the inflammatory response and promote more rapid tissue healing after wound trauma and surgical trauma. This means that brown seaweed broth is recommended after auto collision, sports injuries, bruising falls, muscle and joint damage, and deep tissue cuts, including voluntary surgery. Fucoidan in the pre-surgical patient diet seems to reduce the intensity of blood loss and vascular bed collapse shock during and after surgery. The mechanism for this positive effect is clear.

Biomass for Fuel

More work is necessary to find better methods for the conversion step, biomass to methane, on a large scale, although the bench-scale work already done indicates that net energy can result from bioconversion, with good yields of methane. More engineering research is needed for the design of suitable open-ocean structures that will allow the kelp to survive storms and excessive wave movements and currents. Methane from marine biomass is a long-term project and research and development have been scaled down, probably to be revived when a crisis threatens in natural gas supplies.

Conclusion

The demand for raw material of seaweed based industries leads to extensive and unrestricted commercial harvest of seaweed through the year, there is depletion in the natural stock. The use of seaweed in medicine is not wide spread; the use of seaweed polymer extract in pharmacy, biochemistry is well established. All these and many other uses of seaweeds will demand continuously supply of good quality of raw material. To meet this challenges, it is necessary to develop an appropriate cultivation technology (Anantharaman, 2006).

References

Anantharaman, P., 2006. Resources and conservation of seaweeds. In: *Biodiversity and Conservation of Marine Bioresources*, (Eds.) S. Kannaiyan, T. Balasubramanian, S. Ajmal Khan and K. Venkataraman, pp. 89–106.

Anantharaman, P., Balasubramanian, T. and Thirumaran, G., 2006. Potential value of seaweeds. *National Training Workshop on Seaweed Farming and Processing for Food*, pp. 91–104.

Araki, C., 1978. Some recent studies on the polysaccharides of agarophytes. *Proc. Int. Seaweed Symp.*, 5: 3–17.

Boney, A.D., 1965. Aspects of the biology of the seaweeds of economic importance. *Adv. Mar. Biol.*, 3: 105–253.

Carigie, J.S. and Leigh, C., 1978. Carrageenan and agars. In: *Handbook of Physiological Methods: Physiological and Biochemical Methods*, (Eds.) J.A. Hellebust and J.S. Caraigie. Cambridge University Press, Cambridge, pp 109–131.

Chapman, V.J., 1970. *Seaweeds and their Uses, 2nd edn.* Methuen, London.

Duckworth, M., Hong, K.C. and Yaphe, W., 1971. The agar polysaccharides of *Gracilaria* species. *Carbohydr. Res.*, 18: 1–9.

Gopakumar, K., 1997. Seaweed farming and utilization in the Philippines. *Proc. Workshop National Aquaculture Week*, pp. 64–69.

Izumi, K., 1971. Chemical heterogeneity of the agar from *Gelidium amansii*. *Carbohydr. Res.*, 17: 227–230.

Izumi, K., 1973. Structural analysis of agar-type polysaccharides by NMR spectroscopy. *Biochem. Biophys. Acta.*, 320: 311–317

Ji, Ming Hou, 1990. Processing and extracting of phycocolloids. *Reports on the Regional Workshop on the Culture and Utilization of Seaweeds*, Vol. II: NACA, Bangkok, Thailand, pp. 89–92.

Kaliaperumal, N., 1984. *Report on the Training in Eucheuma Culture at the Marine Sciences Centre*, University of the Philippines, Manila, pp. 1–25.

Kaliaperumal, N., 2000. Seaweed distribution and resources in India. In: *Algalogical Research in India* (Festschirift to Prof. N. Anand). Bishen Singh Mahendra Pal Singh, Dehradun.

Kaliaperumal, N. and Uthirasivam, P., 2001. Commercial scale production of agar from the red algae *Gracilaria edulis* (Gmelin) Silva. *Seaweed Res. Utiln.*, 23(1 and 2): 55–58.

Levring, T., Hope, H.A. and Schmid, O.J., 1969. *Marine Algae: A Survey of Research and Utilization*. Cram, DeGruyter and Co., Hamburg.

Newton, L., 1951. *Seaweed Utilization*. Sampson Low, London.

Pillai, V.K., 1957. Alginic acid from *Sargassum* seaweeds. *Res. Ind.*, 2: 70–71.

Rees, D.A., 1972. Shapely polysaccharides. *Biochem. J.*, 126: 257–273

Sadasivan Pillai, K., 1961. Alginic acid from *Sargassum* seaweeds of Indian coast: Its extraction on a cottage industry basis. Chemical age of India, 12: 425–430.

Viswesra Rao, A. and Mody, I.C., 1964. Extraction of alginic acid and alginates from brown seaweeds. *Indian J. Tech.*, 3(8): 261–262.

Chapter 15

Helminthic Dynamics of Freshwater Clupeid, *Notopterurs notopterus* (Pallas) of Nizamabad District, Andhra Pradesh

☆ *G.S. Jyothirmai, K. Geetha, D. Suneetha Devi, P. Manjusha and Ravi Shankar Piska*

Introduction

The present study of population dynamics of helminth parasites are in the light of some current ecological ideas and concepts. Parasitism is eventually an ecological relationship between population of the species. The essential features of the parasitism are that the parasites are dependent on the hosts physiologically, when compared with the host, the parasites have higher reproductive potential and kill the heavily infected hosts. *Notopterus notopterus* is a freshwater clupeid and its production is reduced due to parasitic diseases. Hence, the present study is undertaken to observe the helminthic infections in the fish.

Materials and Methods

Freshwater clupeid collected from the reservoirs of Nizamabad District, Andhra Pradesh for two years during 2006-07 to 2007-08. 330 and 332 hosts examined for parasitic infection. The live fishes dissected, the organs like stomach and intestine separated and placed in different petri-dishes containing saline water. Various helminthic parasites collected from different biotopes and preserved for identification. The Nematodes identified after the staining. The incidence, density and index of infection, frequency distribution and relative densities studied as per Aruna *et al.* (1986). The statistical analysis were done according to Bailey (1959).

Results

Taxonomical, Biological and Topical Structure of the Parasites

The analysis of the helminth parasites revealed that, in the host *N. notopterus,* the helminth parasites represented by the taxonomical groups like Nematodes represented by four species (Table 15.1). Trematodes and cestodes totally absent.

Table 15.1:Topical Structure of Helminth Grouping of the Fish Host *Notopterus notopterus*

Microhabitat	*Notopterus notopterus*
Stomach	*Spinitectus notopteri* (N)
	Spinitectus thapari (N)
	Spinitectus mastacembali (N)
Intestine	*Metabronema notopteri* (N)

N: Nematode.

It was revealed that in the piscian host. *N. notopterus,* there were 4 geohelminths and no biohelminth recovered (Table 15.2). The Nematode recoverd in the piscian hosts were viviparous involving copepod as the intermediate host. Rerely they may reach the definitive host through small fishes, ingesting these intermediate hosts, or many reach directly.

Table 15.2: Synthesis of Helminth Grouping of Fish Host, *Notopterus notopterus*

Taxonomical Structure		*Biological Structure*		*Topical Structure*	
Namatoda	4	Biohelminths	0	Stomach	3
Trematoda	0	Geohelminths	4	Intestine	1
Total	4	Total	4	Total	4

In *N. notopterus* the topical structure consists of stomach which forms the microhabitat for the parasites like, *Spinitectus notopteri, Spinitectus thapari, Spinitectus mastacembeli* and *Metabronema notopteri. Metabronema notopteri* was occasionally found in the intestine of the host.

Incidence, Intensity, Density and Index of Parasites

The analysis of data revealed that in the host *N. notopterus,* there were two structural groups : *S. notopteri, S. thapari, S mastacembeli* determined as the dominant helminth structure for its remarkable high incidence, high density and high index of infection. *Metabronema notopteri* established as influent helminth structure as they fairly moderate density corresponding to moderate index of infection (Table 15.3 and 15.4).

The investigation of the qualitative structural features revealed the following taxonomic hierarchy of the helminth parasites in piscian host. *N. notopterus.* Nematode species represented by *S. notopteri, S. thapari S. mastacembeli* found to form dominant taxonomic group, the nematode *Metabronema notopteri* formed the influent group (Table 15.4).

Table 15.3: The Mean Incidence, Intensity, Density and Index of Infection of Helminth Parasites of *Notopterus notopterus*

Name of the Parasites	Incidence	Intensity	Density	Index
Spinitectus notopteri	56.9	2.411	1.376	0.867
Spinitectus thapari	87.45	1.988	0.984	0.392
Spinitecuts mastacembeli	42.95	1.777	0.763	0.354
Metabronema notopteri	37.52	1.702	0.642	0.255

Table 15.4: Quantitative Structure of the Helminth Grouping in the Population of the *Notopterus notopterus*

Class of Helminth fauna	Dominant	Influent	Accessory
Nematoda	*Spintecuts notopteri*	*Metabronema notopteri*	–
	Spinitectus thapari		–
	Spinitectus mastacembeli		–

The analysis of the data revealed that the only taxonomic group viz nematode group found to be dominant in the host. In the host, the stomach was most dominant biotope which was densly populated, while the intestine formed accessory topical structure. This similarity and variation may be attributed to several contributory factors of biotic and abiotic nature like habit and habitats of the host, diet and other physiological factors of the host present within the same geographical region.

Overdispersion

Factors like recruitment, seasonal fluctuations, breeding season, age and sex of the host, flow of the parasites and certain meteorological changes control the phenomenon of dispersion over of the parasites. In all parasites the variance was greater than the mean (Table 15.5). The parasites were metabolically and physiologically dependent on the host. Any disturbance in their relation tilts the equilibrium thereby resulting in the damage of the host or parasites or even to both. The degree of overdispersion was significantly varying in dominant, influent and accessory structure of piscian host.

Table 15.5: Comparison of the Mean (X) Variance (S*S) and Dispersion (S*S/X) of the Parasitic Burden in *Notopterus notopterus* during 2006-07 and 2007-08

Name of the Parasite	Hosts Examined		No of Parasites Collected		Mean Values (X)		Variance (S*S)		Dispersion (S*S)/X	
S. notopteri	330	332	452	462	1.369	1.391	3.779	3.774	2.760	2.713
S. thapari	330	332	289	288	0.875	0.869	1.599	1.524	1.827	1.754
S. mastacembeli	330	332	240	260	0.727	0.783	1.380	1.495	1.898	1.909
M. notopteri	330	332	205	221	0.621	0.665	1.368	1.367	2.203	2.056

Relative Density

In *Notopterus notopterus* it was clear from the observed data that the relative density of *Spinitectus notopteri* was higher when compared to other nematodes of the host. It was in the range of 29.69 per cent to 46.43 per cent during the year 2006-07. Similarly, density of this parasite was almost similar which ranged in between 28.36 per cent to 28.33 per cent.

The relative density of *Spinitectus thapari* was moderate, ranged between 20.24 per cent to 28.75 per cent during the year 2007-08. The relative density values were slightly lowered in the next year, ranging in between 17.93 per cent to 35.00 per cent.

The relative density of *Spinitectus mastacembeli* exhibited moderate incidence values. It was the range between 16.11 per cent to 26.15 per cent during the year 2006-07. In the next year it was almost similar which ranged in between 14.65 per cent to 25.00 per cent.

The relative density of *Metabronema notopteri* was remarkably low. In the year 2006-07, it ranged between 13.79 per cent to5 23.44 per cent similarly during the next year 2007-08, the relative density of this parasite was almost similar which ranged between 13.48 per cent to 23.88 per cent (Figures 15.1–15.4).

Discussion

Devi (1995) reported 7 trematodes and 1 nematode in *H. fossilis* and 2 nematodes and 3 trematodes in *Glossogobius giuris*. She reported *Phyllodistomum Indianum, Allocreadium heteropneustes, Orientocreadium*

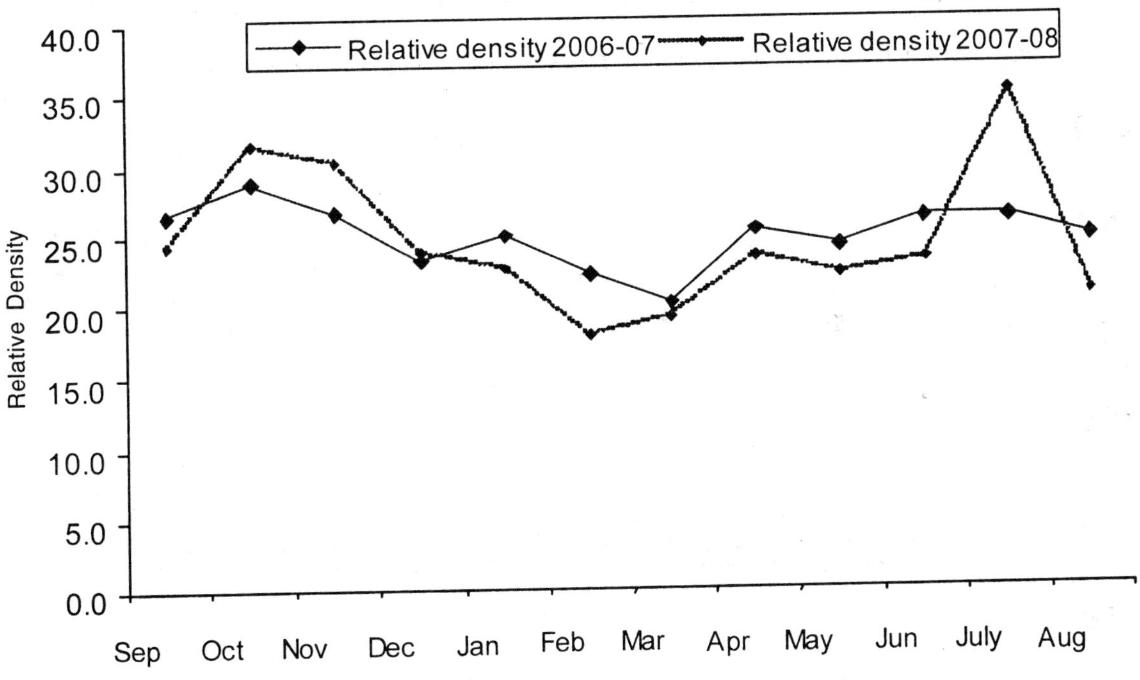

Figure 15.1: Relative Density of *Spinitectus thapari* in *Notopterus notopterus*

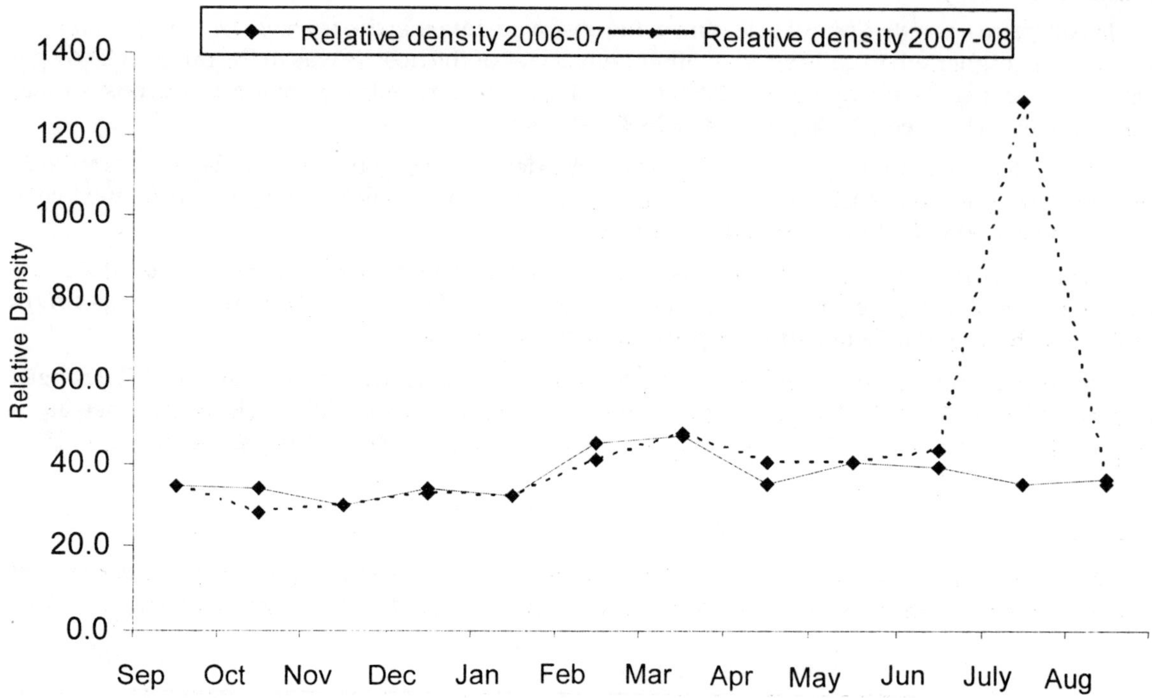

Figure 15.2: Relative Density of *Spinitectus notopteri* in *Notopterus notopterus*

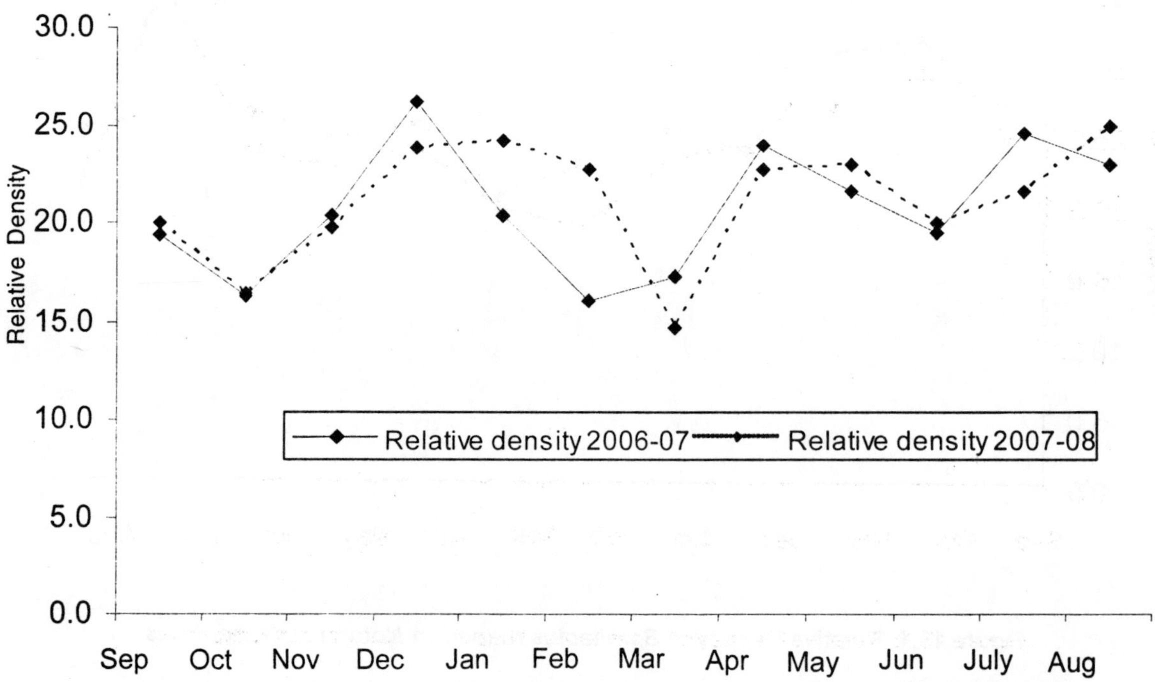

Figure 15.3: Relative Density of *Spinitectus mastacembeli* in *Notopterus notopterus*

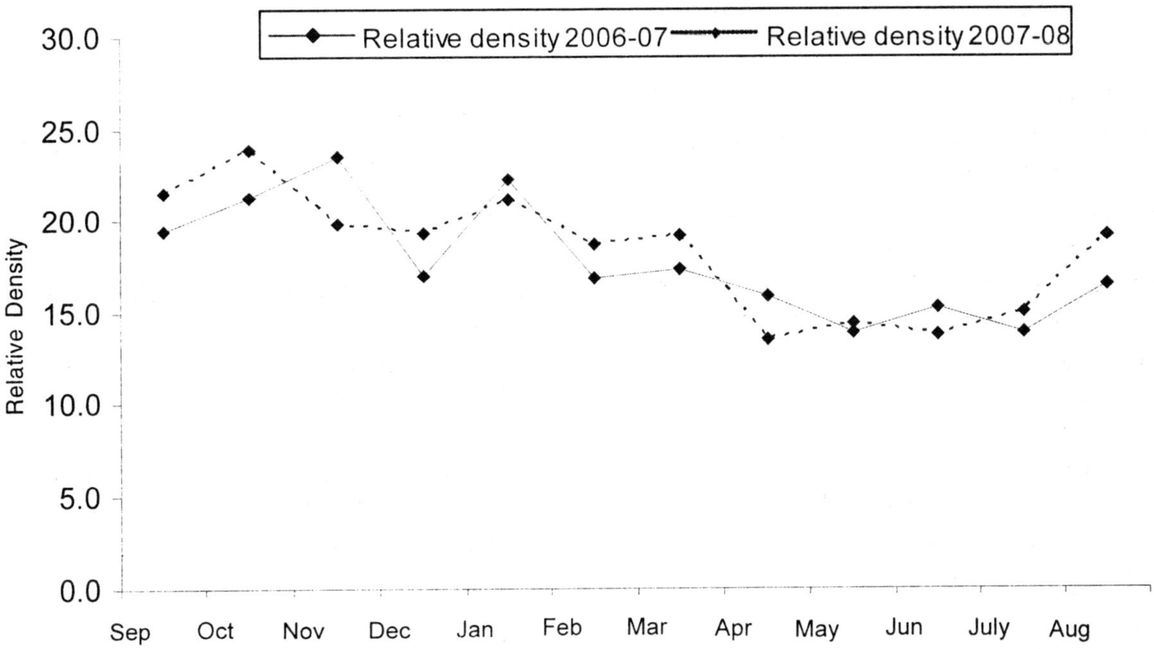

Figure 15.4: Relative Density of *Metabronema notopteri* in *Notopterus notopterus* During the Year 1998–99

dayali, Astiotrema reniferum, Neoplicoelina Saharanpurensis, Cephalogonimus heteropneustes, Masenia fossilis and Procamallanus hetropneustes in H. fossilis and 2 nematodes. And 2 nematodes, *Rhabdochona singhi* and *Philometra,* 3 trematodes, *Allocreadium nicolli, Opegaster beliyai* and *phyllodistomum parorchium* in *G. giuris. Chitikesi* (1997) reported a trematode *Gonorchopsis goppo* and 2 nematodes *Lytoceatus indicus* and *Djumbangia indica* in *Clarius batrachus.* Jyothirmai (2001) reported 3 trematodes *Asymphylodara ritai, Allocreadium thapariand Orientocreadium batrachoides* a cestode, *Proteocephalus ritae* and a nematode, *Procamallanus gubernaculus* in another freshwater cat fish, *Rita rita.*

Devi (1995) reported that geohelminths dominated in *Notopterus and Glassogobius.* These geohelminths found more in stomach. She also reported that there was a reasonably high incidence existed in the helminthic populations and the relative density of nematodes was high. She also reported that the degree of overdispersion depends on the factors like the recruitment and intense flow of the parasites in the hosts, apart from seasonal fluctuations. Chitikesi (1997) reported that the trematodes were domintated in maruf. Jyothirmai (2001) reported that stomach as a biotope was very dominant in *Rita rita* and *Heteropneustes fossilis.*

Rao and Krishna (1984) stated that the relative density of trematodes was very high, moderate in nematodes and very marginal in cestodes in *Rana tigrina.* Aruna *et al.* (1986) reported that the relative density of trematodes was very less. Cestodes was moderate and very high in case nematodes in avian host *Acridotheres tristis.* Rao *et al.* (1988) reported that index of infection was high in nematodes, followed by trematodes and cestodes in a marine fish, *Congrsox talabonoides.* Jain *et al.* (1994) studied the internsity, density and index of infection of *Acanthosenis antspinis* in freshwater cat fish *Mystus seenghala* and concluded that the parasite infection levels were potentially dynamic with durability, regularity and cyclic periodicity.

References

Aruna, V., Ramnivas, Devamma and Rao, V. Rajeshwar, 1986. The relative density of helminth parasites in *Acridotheres tristis. Proc. Indian Acad. Parasito.*, 7(1821): 55–59.

Bailey, N.T.J., 1959. *The Statistical Methods in Biology*. The ELBS and English University Press Ltd., Great Britain.

Chitikesi, R., 1997. Studies on some aspects of Helminth coenosis and host-parasite relationship in *Clarias batrachus. Ph.D. Thesis*, Osmania University, Hyderabad.

Jain, A.K., Devi, Rama and Rao, V. Rajeshwara, 1997. Intensity, density and index of infection of *Acanthosentis antispinis* (Verma Et Datta 1929) in *Mystus seenghala* Sykes, 1839. *Geobios New Reports*, 13: 43–45 (1994).

Jyothirmai, G.S., 2001. Population dynamics of helminthocoenosis of certain freshwater fishes of Nizamabad district, Andhra Pradesh. *Ph.D. Thesis*, Osmania University, Hyderabad.

Rao, S.M., Rao, V. Rajeshwara and Devi, J. Sandhya, 1988. The relatives density of Helminth parasites in *Mystus bleekari. Proc. Indian Acad. Parasitol.*, 9(1): 49–52.

Rao, V.R. and Krishna, G.V. R., 1984. The relative density of helmninth parasites in *Rana tigrina*. In: *Proc. Indian Acad. Prasitol.*, 5(182): 67–70.

Chapter 16

Role of Probiotics in Aquaculture

☆ *M.M. Girkar, A.T. Tandale, S.S. Todkari and B.S. Chaudhari*

Introduction

During the past few years, aquaculture industry has been growing tremendously. Like other industries, this rapid growth has brought with it the problem of environmental pollution. Contamination of coastal waters due to aquaculture is serious concerns among law makers as well as scientists. The coastal environment has been seriously damaged, often resulting in disease outbreaks. Recently, shrimp culture all over the world has been frequently affected by viral and bacterial diseases inflicting huge loss. However, the usage of antibiotics is questioned in recent years, since the pathogenic bacterial strain have a tendency to slowly develop resistance to such drugs.

Now, a day many scientists are trying to use probiotic bacteria in aquaculture to improve water quality by balancing bacterial population in water and reducing pathogenic bacterial load. Scientists are increasingly paying more attention towards ecological aquaculture, and have made considerable headway.

What is Probiotics?

One widely used definition, developed by the World Health Organization and the Food and Agriculture Organization of the United States, is that probiotics are 'live microorganisms, which, when administered in adequate amounts, confer a health benefit on the host.' They are also called 'friendly bacteria' or 'good bacteria.' Probiotics generally includes bacteria such as cyanobacteria, microalgae, fungi etc.

Probiotics are available to consumers mainly in the form of dietary supplements and food. They can be act as complementary and alternative medicine. Probiotics are available in foods and dietary supplements in the form of capsules, tablets, and powders. Examples of foods containing probiotics are yogurt, fermented and unfermented milk, miso, tempeh, some juices and soy beverages. In probiotic foods and supplements, the bacteria may have been present originally or added during preparation.

Most of the, bacteria come from two groups, *Lactobacillus* or *Bifidobacterium*. Within each group, there are different species and within each species, different strains or varieties are available.

Mode of Action of the Probiotic Bacteria

The mechanism of action of the probiotic bacteria has not been studied systematically. According to some recent publications, in the aquaculture the mode of action of the probiotic bacteria may have several aspects.

1. Competitive exclusion of pathogenic bacteria
2. Provide essential source of nutrients that helps in digestion
3. Enhances immune response of cultured organisms
4. Influence on water quality
5. Antiviral effect

Mode of Use of Probiotics

Probiotics are given to the host in following ways:

1. Addition via live food
2. Bathing the organisms
3. Addition of culture water
4. Addition of artificial diet

Probiotics Strains Used in Aquaculture

Probiotics stains commonly used in aquaculture are as follows:

- ☆ *Bacillus acidophilus*
- ☆ *Lactobacillus delbruckii*
- ☆ *L. bulgaricus*
- ☆ *B. subtilis*
- ☆ *B. megaterium*
- ☆ *Acetobacter xylinum*
- ☆ *B. lechiniformis*
- ☆ *Aspergillus niger*
- ☆ *Saccharomyces cerevesiae* (fungus, yeast)
- ☆ *Saccharomyces boulardii* (fungus, yeast)

Conclusion

Based on the previous research results on probiotics we suggest that the use of probiotic bacteria in aquaculture has tremendous scope and the study of the application of probiotics in aquaculture has a glorious future. At present, the probiotics are widely applied in United States of America, Japan, and European countries, Indonesia and Thailand, with commendable results. It is important management tool in the era of fast deteriorating environment. However, the selection of desired probiotic strain needs to be done carefully after understanding the exact mechanism of action.

References

Fuller, R., 1989. Probiotics in man and animals. *J. Appl. Bacterial.*, 66: 365– 378.

Nikoskelainen, S., Salminen, S., Bylund, G., Salminen, S. and Lilius, E.M., 2003. Immune enhancement in rainbow trout (*Oncohynchus mykiss*) by potential probiotic bacteria (*Lactobacillus rhamnosus*). *Fish Shellfish Immunol.*, 15: 443–452.

Olsson, J.C., Westerdahl, A., Conway, P.L. and Kjelleberg, S., 1992. Inestinal colonization potential of turbot (*Scophthalmus maximus*) and dab (*Limanda limanda*) associated bacteria with inhibitory effects against *Vibrio angullarum. Appl. Environ. Microbiol.*, 58: 551–556.

Onarheim, A.M., Wiik, R., Burghardt, J. and Stackerbrandt, E., 1994. Characterization and identification of two *Vibrio* species indeginous to the intestine of fish in cold sea water; description of *Vibrio iliopiscarius* sp. Nov. *Syst. Appl. Micrbiol.*, 17: 370–379.

Queiroz, F. and Boyd, C., 1998. Effects of a bacterial inoculum in channel catfish ponds. *J. World Aquacult. Soc.*, 29: 67–73.

Ringo, E. and Vadstein, O., 1998. Colonization of *Vibrio pelagius* and *Aeromonas caviae* in early developing turbot (*Scophthalmus maximus* L.) larve. *J. Appl. Microbiol.*, 84: 227–233.

Sakai, M., Yoshida, T., Astuta, S., Kobayashi, M., 1995. Enhancment of resistance to vibriosis on rainbow trout, *oncorhynchus mykiss (*walbaum) by oral administration of *Clostridium butyricum* bacteria. *J. Fish Dis.*, 18: 187– 190.

Sakata, T., 1990. Microflora in the digestive tract of fish and shell-fish. In: *Microbiology on Poeciiotherms.* (Ed.) R. Lesel. Elsevier, Amsterdam, pp. 171–176.

Chapter 17

Role of Organics and Probiotics in Shrimp Culture

☆ *R. Saravanan, S. Rajagopal and P. Vijayanand*

Introduction

Aquaculture accounts for about one third of the world's total food supply for food fish. This sector has the potential to become a sustainable practice that can supplement capture fisheries and contribute significantly to feed the worlds exploding populations. However sustainable aquaculture practices over the past few years have lead to serious concern on various coastal biotopes (Primavera 1989; Khor 1995; Yuvaraj 2006; Saravanan 2006). Now a day's International Market poses stringent quality standards for food items. Use of banded antibiotics and other chemicals have caused great concern when their residues appeared in processed products. So, the current study was designed to fulfill the above said lacunae in shrimp culture. The basic principle of organic farming is to encourage natural biological cycles in the production of aquatic organisms, using feed that is not intended or appropriate for human consumption. Organic fertilizers such as manure, cottonseed meal, soybean meal, rice bran, alfalfa meal and other processed grains or hays are used to improve pond productivity in organic farming. The use of animal manure is very effective for stimulating the growth of aquatic plants and animals (Wurts 2001). The use of beneficial bacteria (probiotics) to displace pathogens by competitive processes is being used in organic farming as a better remedy than administering antibiotics. The global market (U.S., Europe and Germany) for the organic food is booming, worth approximately US$ 20 billion in the year 2000 (Lockwood 2000; Ramachandran and Sathiadas 2005). Keeping this in mind the present study is aimed to study the use of organics and probiotics in culture of black tiger shrimp *Penaeus monodon*.

Materials and Methods

The study was carried out in a shrimp farm situated on the southern bank of Uppanar estuary at Thirumulaivasal Tamil Nadu, India. Two ponds were selected for the present study, a control pond and an experimental pond. Both the ponds had a water spread area of 0.6ha. Soil culture was done by applying shell lime at 500 kg for both ponds. Disinfected water was pumped from the reservoir to both the ponds. Organic fertilization (rice bran, cow dung, yeast and a blend of probiotic bacteria) was done in the experimental pond were inoculated for plankton production, whereas in the control pond inorganic fertilizers were added in the ratio of 10: 2 (N: P).

Healthy (negative-Polymerase Chain Reaction (PCR)) *Penaeus monodon* seeds were stocked at a density of 6/m^2 in both ponds. Survival rate was estimated using survival pens (happa nets). Blind feeding was done for the first 30 days. Later the feeding was adjusted according to the check tray observation. Feeding was done by using CP feed (Charoen Pokhpand aquaculture India Pvt.Ltd, Chennai, India) and the feeding schedule was based on the company's feed chart. Sampling was done in the ponds every fortnight during early hours of the day with a cast net. Healthiness, survival rate, average body weight (ABW) and average daily growth (ADG) of the animals was recorded. Regular water exchange was done only to the control pond, and for the experimental pond only topping up of water was done. Commercially available probiotics NS Super SPO (Nu Genes Technologies, Alexander Avenue, Inc.) was used as water probiotic and Super PS (CP aquaculture India Pvt. Ltd.) as soil probiotic.

Water quality parameters were regularly monitored at a regular interval of 15 days by using standard water analysis kits. Nutrients (Ammonia, Nitrite, Nitrate, Total phosphate, Silicate) were estimated following the method described by Strickland and Parsons (1972). The parameters were statistically treated by using ANOVA (Table 17.1 to 17.5). The Total Heterotrophic Bacterial (THB) populations were analyzed in water and sediment samples of two ponds. For studying the THB population, dehydrated bacteriological media, Zobell's 2216 (Himedia Laboratories Private Limited, Mumbai, India) was dissolved in 50 per cent seawater and sterilized by autoclaving at 15lb preassure for 15 minutes. THB population was estimated by following the pour plate technique.

Results

Water quality parameters were recorded in both the ponds (control and experimental ponds). Temperature does not show much difference and it range from 27.14°C to 30.36°C in the experimental pond where as in the control pond it was between 28.66°C to 30.86°C during the culture period. Salinity levels varied between 32.9 to 37.4ppt in the experimental pond. In the control pond it varied from 28 to 39ppt. pH value in the experimental pond ranged from 7.38 to 7.98 and in the control pond the values ranged between 7.5 to 7.78. The level of dissolved oxygen (DO) was more in experimental pond, DO concentration varied from 5.13 to 5.6 mg/l in the experimental pond and in the control pond it ranged between 4.98 to 5.39 mg/l. Transparency of water in the experimental pond is low and it ranged from 25 to 37cms.In the control pond it ranges from 29 to 41cms. The ammonia concentration was low in experimental pond, varied from 0.338 to 0.613 ppm in and in the control pond it ranged between 0.278 and 1.124 ppm. Nitrite concentration range from 0.0019 to 0.0067 in the experimental pond and in control pond it range from 0.0032 to 0.0171. The level of nitrite is low in experimental pond when compare to control pond. Nitrate concentration was high in experimental pond and it varied from 0.0034 to 0.0171 ppm. In the control pond it varied from 0.0024 to 0.0126. In the experimental ponds the total phosphate concentration varied from 0.0032 to 0.0176 ppm. The concentration of total phosphate in the control pond is low and it ranged between 0.0029 to 0.0126 ppm. The silicate concentration in the experimental pond range from 0.0013 to 0.0075 ppm and in control pond it range from 0.0012 to 0.0047 ppm. The probiotics plays a major role to increase the nutrients level in the experimental pond. Results of the physico-chemical parameters of the ponds were given in Table 17.1.

Table 17.1: Water Quality Parameters in Both Experimental and Control Ponds

Water Quality Parameters	*Experimental Pond (Mean ± Standard Deviation)*	*Control Pond (Mean ± Standard Deviation)*
Temperature (°C)	27.14±1.79 to 30±0.89	28.66±1.02 to 30.86±1.09
Salinity (ppt)	32.9±2.42 to 37.4±1.43	28±1.66 to 39±1.34
pH	7.38±0.18 to 7.98±0.19	7.5±0.26 to 7.78±0.29
Dissolved Oxygen (mg/l)	5.13±0.21 to 5.6±0.18	4.98±0.18 to 5.39±0.32
Transparency (cm)	25.0±1.5 to 37.0±1.00	29.0±1.14 to 41.0±1.33
Ammonia (ppm)	0.338±0.019 to 0.613±0.073	0.278±0.07 to 1.124±0.016
Nitrite (ppm)	0.0019±0.0004 to 0.0067±0.0005	0.0032±0.0004 to 0.0171±0.0002
Nitrate (ppm)	0.0034±0.0003 to 0.0171±0.0002	0.0024±0.0005 to 0.0126±0.0002
Total Phosphate (ppm)	0.0032±0.0003 to 0.0176±0.0001	0.0029±0.0005 to 0.0126±0.0006
Silicate (ppm)	0.0013±0.0003 to 0.0075±0.0002	0.0012±0.0006 to 0.0047±0.0003

In the water samples the THB population ranged from 4.1×10^3 to 9.3×10^5 CFU/ml in the experimental pond and in control pond, the microbial population was range from 2.4×10^3 to 8.3×10^6 CFU/ml. In sediment, the THB load was found to be between 3.9×10^4 and 9.3×10^8 CFU/g in the experimental pond and in the control pond the load varied from 3.7×10^4 to 9.2×10^7CFU/g. THB was found to be higher in the sediments than in water and THB seems to higher in experimental pond than in control pond. The results have been shown in Table 17.2 and 17.3. The daily growth rate was high in the experimental ponds (0.23 g), and in control pond (0.22 g). The percentage of survival was higher in experimental ponds (75 per cent) when compared to the control pond (65 per cent) Results have been displayed in Table 17.4 and 17.5.

Table 17.2: THB Population in Water and Sediment Sample in the Experimental Pond

Days of Culture	*Water (CFU/ml)*	*Sediment (CFU/gm)*
15	4.1×10^3	3.9×10^4
30	6.3×10^3	5.6×10^4
45	5.1×10^4	7.3×10^5
60	9.5×10^4	1.4×10^6
75	3.6×10^5	6.3×10^6
90	5.1×10^5	5.8×10^7
105	6.8×10^5	9.3×10^8
120	9.3×10^5	6.6×10^7

Discussion

In the present study two adjacent ponds were selected considering one pond as experimental (organic) and second as control. Different farming methods were employed and monitored for a culture period of four months. Fertilization was done with organic manure in the experimental pond and super phosphate and urea in control pond. Organic manure promotes the growth of zooplankton and phytoplankton as well as other invertebrates and pond micro-organisms. An abundance of natural food organisms in fertilized ponds minimize stress and mortality in shrimp post larvae (Wurts,2001). The same condition was observed in the present study. Organic fertilizers also stimulates the growth of decomposers such as bacteria and fungi which break down the toxic waste and there by stabilize the water quality, increasing survival and enhancing the growth of the shrimps (Yuvaraj, 2006).

Table 17.3: THB Population in Water and Sediment Samples in the Control Pond

Days of Culture	*Water (CFU/ml)*	*Sediment (CFU/gm)*
15	2.4×10^3	3.7×10^4
30	5.9×10^3	4.7×10^4
45	1.2×10^4	5.8×10^5
60	7.2×10^4	3.7×10^6
75	9.1×10^4	5.8×10^6
90	2.3×10^5	1.6×10^7
105	1.5×10^6	4.7×10^7
120	8.3×10^6	9.2×10^7

Table 17.4: Growth and Survival of Shrimps in Experimental Pond

Days of Culture	*Survival No.*	*ABW (gm)*	*Biomass (kg)*	*ADG*
30	91	3.9	177.45	0.13
45	87	4.9	213.15	0.10
60	85	7.3	310.25	0.12
75	81	14	567.0	0.18
90	79	19.1	754.45	0.21
105	77	24.7	950.95	0.23
120	75	28.6	1072.5	0.23

Table 17.5: Growth and Survival of Shrimps in Control Pond

Days of Culture	*Survival No.*	*ABW (gm)*	*Biomass (kg)*	*ADG*
30	85	3.3	140.2	0.11
45	80	4.1	164.0	0.09
60	76	6.6	250.8	0.11
75	73	12.8	467.2	0.17
90	70	17.4	609.0	0.19
105	68	23.3	792.2	0.22
120	65	26.5	861.2	0.22

Shrimps were stocked in both the ponds at $6/m^2$ after proper acclimatization. Production of shrimp using organic methods should encourage animal's natural health and behaviour that are identical to conditions they would have in the wild. Hence stocking density was reduced providing surplus moving space. Growing shrimps in both the ponds were fed using the same commercially available feed. According to Organic Standards, antibiotics, synthetic growth agents, antioxidants, appetite stimulations, pure amino acids and synthetic dyes tiffs are prohibited in feed diets. The selected feed is devoid of the above factors and was prepared from naturally occurring components.

Water exchange is important to maintain the water quality and periodical removal of waste in the culture pond. In the present study, water exchange was not carried out in the experimental pond but periodical pumping of water was done to maintain the water level. In control pond, periodical water exchange was carried out. By avoiding water exchange in experimental ponds, effluent discharge is minimized there by reducing nutrient and biological pollution in surrounding water bodies. Production of shrimps in zero water exchange systems have been demonstrated earlier and proved to have increased survival and mean weight of cultured organisms (Samocha *et al.*, 1998). Experimental pond found to be extremely efficient in assimilating organic matter and nutrients. Huge fluctuations of salinity in the control pond are because of the regular water exchange.

Probiotics refers to a bacterial supplement of single or mixed culture of selected bacterial strains (Prabhu *et al.*, 1999). In the experimental pond a supplement of nitrifying bacteria were added at regular intervals to improve the pond water and sediment quality. Probiotics have the potential benefits to enhancing the decomposition of organic matter, reducing nitrogen and phosphorous concentrations, better algal blooms, greater availability of dissolved oxygen, control of ammonia, nitrite and hydrogen sulfide, lower incidence of disease, greater survival and greater shrimp production (Boyd and Gross, 1998). Plankton productivity is more because of organic manure in experimental pond. Application of commercially available water probiotics, NS Super SPO (Nu Genes Technologies, Alexander Avenue, Inc.) and soil probiotic Super PS (CP aquaculture India Pvt. Ltd.) results coincides with the above findings.

Good water quality plays a vital role in aquaculture production. Being the media where it lives, water plays an important role in the survival and growth of shrimp (Rajagopal *et al.*, 1995). Pond water affects both the abundance and species composition of micro flora in shrimp gut (Moss *et al.*, 2000). It also stimulates digestive enzyme activity in shrimps and enhances the immune response by stimulating agglutination (Primavera *et al.*, 2000). The quality of water during culture will deteriorate mainly due to the accumulation of unutilized feed, animal excreta pond decayed plankton. Generally organisms are in a state of balance between potential disease causing organisms and their environment. Any change in this equilibrium can affect the growth and survival of organisms as they become vulnerable to diseases due to the stress. Water quality parameters such as salinity, temperature, pH, DO, transparency, ammonia, nitrate, total phosphate and silicate affect the growth and production of shrimps in aquaculture systems. Conservative parameters such as temperature and salinity are environmental parameters which are unaffected by biological process (Prabhu *et al.*, 1999). Other parameters were found to vary with the progress of culture in both the ponds during the study period.

Micro-organisms play a critical role in maintaining acceptable water quality by providing oxygen, converting toxic metabolites to less toxic forms and oxidizing organic matter. Uneaten feed reaches the pond bottom to be decomposed by microbes and converted to inorganic nutrients such as ammonia, phosphate and CO_2. They also have profound and direct impact on shrimp growth and survival (Leber and Pruder 1988; Moss *et al.*, 1992; Yuvaraj 2006; Saravanan, 2006). The assimilative capacity of microbes to process and recycle organic matter and remove the toxic metabolites from the culture environment is determined largely by the quality and quantity of exogenous feed supplied to the system (Brune and Drapcho 1991). In general bacterial productivity was high in sediment than in water. In the present study the THB load is found to be higher in experimental pond than in control pond. Thus the increase in bacterial population can be attributed to the multiplication of beneficial bacteria in the culture pond.

The growth of shrimps in experimental pond was encouraging compared to the control pond. Shrimps attained an average growth of 28.6g in 120 days whereas the growth of the shrimps in control pond was 26.5g. Increase in the growth of the organism in probiotics medium is attributed to external nutrients supplied by probiotics strains besides enzymes that enhance the digestive process. Addition of probiotics changes the bacterial composition of rearing medium, colonization of gut, modification of gut flora (Fuller, 1977) and alteration in host metabolism (Deeth, 1984).

Organically farmed seafood is gaining momentum and have great potential for allowing consumers to know that the fish they purchased were farmed in environmentally friendly way this in turn provide responsible farmers to receive a premium price for their product (White *et al.*, 2004) to a tune of 20–30 per cent higher than the conventional product (Hoe, 2001). Considering the various water quality parameters, build up of bacterial populations and the growth performance of shrimps, it can be concluded that this farming practice involving the use of naturally available organics is successful and economically viable.

References

Boyd, C.E., 1982. *Water Quality Management for Pond Fish Culture*. Elsevier Amsterdam, p. 318.

Boyd, C.E. and Gross, A., 1998. Use of probiotics for improving soil and water quality in aquaculture ponds. In: *Advances in Shrimp Biotechnology*, (Ed.) T.W. Hegel. National Centre for Genetic Engineering and Biotechnology, pp. 101–106.

Brune, D.E. and Drapcho, C.M., 1991. Fed pond aquaculture. In: *Aquaculture Systems Engineering*, (Ed.) P. Giovannini. St. Joseph, MI USA, pp. 15–33.

Deeth, H.C., 1984. Yogurt and culture product. *Australian Journal of Dairy Technology*, 39: 111–113.

Fuller, R., 1977. The importance of *Lactobacilli* in maintaining normal microbial balance in the crop. *Poultry Science*, pp. 85–94.

Hoe, T.D., 2001. Shrimp industry and production of organic shrimp in Vietnam. *Aqua International*, 9(6): 11.

International Federation of Organic Agriculture Movements (IFOAM), 2000. Basic Standards for Organic Production and Processing, Available at: www.ifoam. Org/standard/basics.html#11.

Khor, M., 1995. The aquaculture disaster. *Third World Resurgence*, 59: 8–10.

Leber, K.M. and Pruder, G.D., 1988. Using experimental microcosms in shrimp research: The growth enhancing effect of shrimp pond water. *Journal of the World Aquaculture Soc.*, 19: 197–203.

Lockwood, G.M., 2000. Organic fish: A major market opportunity. *Aquaculture Magazine*, 26(6): 24–28.

Moss, S.M., Prider, G.D., Leber, K.M. and Wyban, J.A., 1992. Relative enhancement of *Penaeus vannamei* growth by selected fractions of shrimp pond water. *Aquaculture*, 101: 229–239.

Moss, S.M., Lea Master, B.R. and Sweeney, J.N., 2000. Relative abundance and species composition of Gram–negative, aerobic bacteria associated with the gut of juvenile white shrimp *Litopenaeus vannamei* reared in oligotrophic well water and eutrophic pond water. *Journal of the World Aquaculture Society*, 31: 255–263.

Prabhu, N.M., Nazar, A.R., Rajagopal, S. and Khan, S. Ajmal, 1999.Use of probiotics in water quality management during shrimp culture. *J. Aqua. Trop.*, 14(3): 227–236.

Primavera, J.H., 1989. The social, ecological and economic implications of intensive shrimp farm. *Asian Aquaculture,* 11: 1–6.

Primavera, J.H., Estepa, F.D.P., and Lebata, J.L., 1998. Morphometric relationship of length and weight of giant prawn *Penaeus monodon* according to life stage, sex and source. *Aquaculture,* 164: 67–75.

Primavera, J.H., Lea Master, B.R. and Boss, S.M., 2001. Environmental induction of *Vibratos harveyi* serum agglutinins in the Pacific white shrimp *Litopenaeus vannamei.* In: *8th Congress of the International Society of Developmental and Comparitive Immunology*, p. 24.

Rajagopal, S., Srinivasan, M. and Khan, S. Ajmal, 1995. Problems in culturing black tiger shrimp (*Penaeus monodon*) the semi–intensive way: An Indian experience. *NAGA, The ICLARM Quarterly,* 18: 29–30.

Ramachandran, C. and Sathiadhas, R., 2005. Greening the Pink Gold: A perspective on the economic potential and market trade prospects of organic aqua-products in India. *Seafood Export Journal,* 35: 15–25.

Samocha, T.M., Lawrence, A.L., Horowitz, A. and Horowitz, S., 1998. Commercial bacterial supplement its potential use in the production of marine shrimp under no water exchange. In: *Proceedings of the First Latin American Shrimp Culture Congress and Exhibition*, (Ed.) D.E. Jory. October 1998, Panama City, Panama.

Saravanan, R., 2006. Study on use of organics in shrimp culture (*Penaeus monodon*). *M.Phil Thesis,* Centre of Advanced Study in Marine Biology, Annamalai University, India.

Schwartz, M.F. and Boyd, C.E., 1994a. Effluent during harvest of channel catfish from water ponds. *The Progressive Fish-Culturist*, 56: 1–5.

White, K., Brendran, O' Neill and Tzankova, Z., 2004. A sea web aquaculture clearinghouse report. www.AquacultureClearinghouse.org, pp. 1–15.

Wurts, W.A., 2001. Organic fertilization in culture ponds. *World Aquaculture,* 35(2): 64–65.

Yuvaraj, D., 2005. Studies on some aspects of shrimp culture (*Penaeus monodon*). *Ph.D. Thesis,* Centre of Advanced Study in Marine Biology, Annamalai University, India.

Chapter 18

Prospectus of Magur Culture

☆ *M.M. Girkar, S.S. Todkari, S.B. Satam and A.T. Tandale*

Introduction

The most important aquaculture species of Asian catfishes *Clarias batrachus* (family Claridae) have consumer preference and their culture systems are yet to be established in many countries of Asia. Among the catfishes, *Clarias batrachus* is of utmost importance owing to its taste, medicinal and high market value. This fish popularly known as magur, is an air breathing fish and well adapted to adverse ecological conditions. They normally inhabit swamps, marshy and derelict waters. These water bodies are usually shallow with heavy silt of decaying vegetation and organic load with poor nutrient release. Besides, these water-bodies have low pH, oxygen and primary productivity. On the contrary they have high carbon dioxide, hydrogen sulphide, methane and free ammonia, and this type of adverse environment is quite insensitive to the above air breathing slow growing, hardy omnivorous fishes. CIFA, Barrackpore has standardized the following technologies on the breeding, seed production, larval rearing and grow out system of this precious species.

The suitability of this species for pond culture is based on the following biological criteria (Hogendoorn, 1979):

1. It matures and is relatively easy to reproduce in captivity
2. It can grow fast and efficiently
3. It supports high population densities
4. It is hardy in nature and it can tolerate wide range of environmental conditions.
5. To adapt to fresh and brackish water.

Distribution

Clarias gariepinus locally called magur, is an indigenous species in Africa where it is widely distributed. It naturally inhabits tropical swamps, lakes, rivers and floodplains some of which are subject to seasonal drying. In recent years the species has been introduced in Europe, Asia and South-America. The species easily adapts to environments, where the water temperature is higher than 20°C.

Identification Characteristics

1. Its body is elongated cylindrical and anguilliform shape.
2. Magur has a scaleless slimy skin, which is darkly pigmented in the dorsal and lateral parts of the body.
3. The head is flattened, highly ossified, the skull bones (above and on the sides) forming a casque.
4. The length of head is 30–35 per cent of body length.
5. Around the mouth, 4 pairs of barbels can be distinguished (nasal, maxillary, the longest and most mobile, outer mandibular and inner mandibular). The magur can move the maxillary barbels independently of its mouth. The barbels serve as tentacles. Close to the nasal barbels, two olfactory organs are located. Magur recognizes its prey mainly by touch and smell. This is of relevance during feeding at night and in highly turbid or muddy waters where visibility is less.
6. Mouth is wide able to feed on a variety of food items ranging from minute zooplankton to large fish. They are able to suck benthos from the bottom, tear pieces of cadavers with the small teeth on its jaws and to swallow prey such as a whole fish.

Foods and Feeding Habit

Micha (1973) considered magur as an omnivorous fish with a high tendency for predation. Different kinds of food items were found by different authors in the stomach of the African magur captured from natural waters. The food items reported are aquatic and terrestrial insects, fish, molluscs, fruits, diatoms, arachnids, plant debris, seeds, detritus, bird eggs, young birds, droppings of poultry and zooplankton. In zooplankton-rich fish pond, African magur often join other carp species to graze zooplankton on the surface. Many authors concluded that the African magur is an omnivorous slow moving predatory fish which feeds on a wide variety of food items from zooplankton to fishes of half of its own length (Janssen, 1987).

Culture Techniques

General Considerations

Various production conditions can be found in Bangladesh. It seems that magur can be adopted to these wide range of possibilities. However, intensity of production depends on the following parameters:

Pond Size

A pond of 0.75–1.5 ha is ideal for production of magur. Most of the small ponds around the farm are suitable for magur stocking after some modifications. But the large, more than 1 acre undrainable ponds are not recommended for stocking with African magur. For large scale production properly designed ponds should be constructed. Depth of pond for grow out should be within the range of 1.5–2.0 m.

Availability of Fingerlings

Year-round, multi-crop culture of African magur largely depends on availability of fingerlings throughout the year. Under the local conditions of Bangladesh, it is now possible to breed magur for at least eight months in a year.

Availability of Inputs

The input demand increases exponentially with the intensity of production. But in small scale culture where magur, is a small component in polyculture, the input cost is little. With the increase in intensity of production the input costs rise sharply.

Availability of Feed Ingredients

For African magur culture, the protein feed supply is the main limiting factor. One should find out first where the slaughterhouse byproducts or other cheap animal protein sources are available on a daily basis and in adequate quantities.

Marketing

The present market system (mainly in rural areas) is favourable to small scale operations. When the production of African magur is specialized and intensified, the market channel should be found out before stocking of fingerlings. For commercial production of magur, detailed market study should be undertaken on such operations.

Preparing Pond for Magur Production

Before deciding to construct a new pond for magur culture the followings should be checked:

1. The area should be free from flood free even at the highest flood level
2. Water retention capacity of the soil is good.
3. Dig a 1–1.2 m hole on the proposed site and fill the hole with water. Next day fill again to make up the water loss by seepage. After 24 hours the water level and the wall should be checked. If the water level remain more or less same the soil is suitable for magur pond.
4. While making embankments, roots, leaves and other organic particles should be removed from the soil, and the soil should be compacted well. Local traditions and low initial input demands support the simple undrainable ponds.
5. The pond bottom should be prepared flat with slight slope towards the outlet or the usual pumping place. Grass should be planted on the top and the slopes of dykes to reduce erosion.

Eradication of Predator and Weed Fishes

Before stocking fingerlings in the pond it is necessary to eradicate the unwanted fish from the pond. Predator and weed fish tend to decrease yield considerably. Repeated netting may not be sufficient. It is, therefore, necessary to apply fish toxicants. Some of the efficient toxicants listed below can be used:

Efficient Toxicants	*Concentration of Dose (ppm)*
Rotenone	2–3 ppm
Phostoxin	0.2 ppm (Toxicity lasts for around 10–15 days)
Tea seed cake	75–100 ppm (The toxicity lasts for about 10–12 days)
Mahua oil cake	250 ppm (The toxicity lasts for about 10–15 days)
Bleaching Powder	30 ppm (Its toxicity lasts for about 7–8 days)

Liming

Liming of a fish pond is highly recommended if the soil is not alkaline. Lime neutralizes soil acidity and creates a buffer system to prevent marked diurnal changes of the water from acidic to alkaline conditions. Apart from other advantages, this buffering action of calcium is the most important. Lime treatment for ponds should be done before initial manuring as follows:

Soil pH	*Soil Type*	*Quantity of Lime ($CaCO_3$) (Kg/ha)*
4.0–4.9	Highly acidic	270
5.0–6.4	Moderately acidic	140
6.5–7.4	Near neutral	70
7.5–8.4	Mildly alkaline	30
8.5–9.5	Highly alkaline	No liming

Initial Fertilization

A well prepared zooplankton rich water is the best starter for magur fingerlings. The magur fingerlings try to supplement its higher protein requirement from natural sources. At stocking when the size is 2–3 cm, the bigger size zooplankton *e.g. Cladocera* sp., and *Diaptomus* is the cheapest and best food source for fingerlings.

In different types of magur culture in earthen ponds considerable quantities of nutrient elements are removed from the pond ecosystem through fish production. Therefore the fertilizer requirements vary among different soil productivity levels and the intensity of production. Organic manuring besides being important as means of adding the nutrients is also equally important for improving the soil texture. During the shortage of organic manures, application of inorganic fertilizers is recommended.

Manures/Fertilizers	*Quantity (kg/ha)*
Chicken manure (dry)	250
or Cow dung	400
Urea	20
Triple/ Single Super Phosphate	7

Induced Breeding Techniques

Usually this species attains maturity at the age of one year and breeds during June-August. During breeding season male and female magur are collected from brood stock pond, and kept separately in plastic tubs containing water. Males and females are distinguished by their secondary sexual characters. In females the abdomen is gravid, vent is reddish colored, genital papilla is round and button shaped. In male genital papilla is elongated and pointed. The brood fish of 100-150 g size can be used for successful spawning. For the purpose, they should be well fed with feed containing 30 per cent protein diet daily, at least 3 months prior to breeding season. The females are either induced through hormone therapy using carp pituitary extract (30-40 mg/kg), HCG (4000 IU/kg) and Ovaprim or Ovatide (0.6-0.8 ml/kg) or through environmental manipulation in a controlled system for spontaneous spawning. The injected fishes are to be stripped after 14-17 hr for releasing eggs and fertilized artificially with sperm suspension made with mascerating the testes of the male fish. Shining and brown coloured beads like eggs are considered as good eggs, while white coloured eggs are of bad quality. A 150 g female lays around 7,000-9,000 eggs.

The flow-through hatchery consist a row of small plastic tubs of 12 cm diameter, 6 cm height placed on a cement platform and are provided with flow-through water system. The water supply is provided from an over head tank through a common pipe to all the tubs with individual control taps. Each tub is having the provision of an outlet at a height of about 4 cm. The fertilized eggs are uniformly distributed in the incubation tubs.

For large scale hatching, an improvised hatchery system has also been developed, which consists of a circular tank having 2 m diameter with inlets at a height of 15 cm at an angle of 45°. A feeble inflow of water is maintained. The eggs are uniformly spread for hatching. The system can accommodate 1 lakh fertilized eggs at a single operation with 60-80 per cent hatching rate. After hatching, the larvae are collected by siphoning from the bottom of the hatching tank.

Ideal temperature for hatching is between 27-30°C. Hatching takes place within 24-26 hours. The tubs from the flow-through hatchery are washed properly to make the larvae free from egg shell. The magur larvae with yolk sac measures about 4-5 mm in length.

Larval Rearing

The larvae are reared in indoor rearing tanks. There is no necessity to provide feed during first three days as yolksac in larvae serves as the stored feed. After yolksac absorption, the larvae are fed with either live plankton or *Artemia* nauplii. It is important to provide good environment to the larvae. For the purpose the indoor rearing tanks are provided with continuous aeration and water exchange facilities. There is a chance of mortality and poor growth of larvae due to poor environment and high stocking density during in-door rearing phase. A stocking density of 2000-3000 nos/m^2 is considered to be optimum for better growth and survival in indoor condition. The larvae grow to 10-20 mm or30-40 mg fry during 12-14 days of rearing. After a maximum of 14 days rearing in the indoor, they should be transferred to out-door rearing tanks for fingerling production.

The advanced fry are further reared in cemented cisterns or earthen ponds for fingerling production. Generally the advanced fry reared in pond condition do not show good survival due to natural mortality or predation as in this stage the fish does not have much capacity to escape from predators. Therefore, small cemented tanks of 10-20 m^2 size are required for better survival and easy management. The tanks are prepared with soil base and manured like carp nursery ponds. The tanks are inoculated with plankton and the advanced fry are stocked after 6-7 days of preparation. The fry grows to a size of 0.8-1 g during 30 days of rearing when fed with laboratory made feed containing 30 per cent protein and at a density of 200-300 nos/m^2.

Stocking of Pond

The fry and fingerling can be transported from nurseries when the pond is ready for stocking. The fry is ready for stocking when the arborescent organs are developed. It is completed after about two weeks of age. It is easy to identify by the special behavior of air gulping. The small magur fry vertically swims up to the surface to breathe air and quickly goes down. Size of fingerlings should be equal. Size differences in magur naturally occur from the beginning of rearing. Fingerlings of equal size should be sorted out from the stock. The fingerlings may be kept in a net for a short time, the mesh size of which will allow the small ones swim out. The bigger size fingerlings should be picked up by hand during counting. The initial size differences will be even more during the grow-out period and that difference creates problems like cannibalism.

The following signs should be help in identifying the healthy from the non-healthy fingerlings

Sl.No.	Healthy magur fingerlings	Non-Healthy magur fingerlings
1.	Quick moving and gathering on the bottom of the net	The fingerlings hang vertically on the surface with weak response to disturbances
2.	Fast swimming up and down when gulping air	Slugish swimming up and down when gulping air
3.	Unbroken barbels and fins	Barbels and fins are broken
4.	Skin is uniformly dark-grey colour covered with mucus	Skin is dull dark-grey colour covered with no mucus

For short distance transportation magur fry can be transported in plastic buckets with led. For long distance transportation, plastic bags with oxygen should be used. For this, the fingerlings should be held in a hapa for 12 hours to get their guts cleared. Otherwise, decaying faeces in the water inside the plastic bag produces ammonia which may cause high mortality during long distance transportation. Before releasing in the pond, the fingerlings should be acclimatized to the temperature of the pond water. This can be done by slowly adding pond water into the water containing the fingerlings inside the plastic bag.

The stocking rate of magur and the supplementary carp species will vary according to (i) the availability of feed for magur and (ii) the pond conditions, especially on the quantity and the quantity of natural feed present in the pond. However, the following stocking rate is recommended for semi-intensive magur dominated polyculture system.

Grow Out Culture

The earthen ponds or stone pitched ponds or cemented tanks are suitable for grow-out culture of magur. Generally high density of 50,000-70,000/ha is recommended for culture of this species. Bigger sized fish (5-10 g) shows good survival and growth during culture. The fishes are fed at the rate of 3-5 per cent of their body weight with pelleted feed in the feeding basket placed in different places of the pond.

Since they are air breather, they normally come up to the water surface for atmospheric oxygen. This kind of habit attracts birds for predation. Therefore, it is required to cover the ponds with net to protect the fishes. The fishes attain a marketable size of 100-120 g during a period of 7-8 months. Harvesting of magur is done by dewatering the pond completely and picking them manually from the culture ponds. Productions of 3-4 tonnes are achieved from one hectare of water area.

Feed

In semi-intensive production level the growth of magur is highly dependent on the quality and quantity of supplementary feed. As the stocking rate in the pond increases the feed requirement also increases. Protein is the most expensive component of the magur supplementary feed and is the most important nutrient in the diet. Protein sources should be first identified from the ingredients available within reasonable distance. In semi-intensive pond culture most of the protein demand must be met through the supplementary feed. Smaller part is gathered by the fish from the well prepared pond water. Only a small part of the protein requirement may be met from the natural feed available in the pond. The basic level of nutrient requirement of African magur is given in following Table.

Nutrient Requirement of African Magur (Modified after Janssen 1987)

Nutrients	*Per cent of Dry Matter*
Protein	30–35
Digestible energy	2,500–3,500 kcal/kg
Ca	0.5–1.8
P (available)	0.5–1.0

Common Diseases in Magur

In well managed culture conditions African magur is found to be more or less resistant to fish pathogens *viz.*, virus, bacteria and parasites. Poor water quality, infected feed and rough handling can make the fish weaker and make them susceptible to diseases.

Stressed or infected fish can be easily recognized by some abnormal behaviors, such as poor appetite and abnormal swimming behaviour, (staying in vertical position on the pond surface). In addition, some clinical symptoms such as mutilated barbels or fins, white or red-brown spots on the skin, pop-eyes etc could be observed. The health conditions of magur should be monitored daily, particularly during feeding when magur frequently come to the surface water. For diagnosing bacterial, fungal and parasitic diseases, squash preparations of the skin, the gill filaments, intestines etc. have to be made and examined.

Bacterial Infections

Columnaris Disease

Caused by: Flexibacter columnaris

Symptoms: The fish remain in vertical position at the water surface. Big white spots like lesions on the body without mucus. Fins are broken.

Treatment: Antibiotics such as Chloramphenicol, Terramycin or Oxytetracycline can be added to the feed. Dose : 5 to 7.5 g/100 kg fish/day for 5–10 days.

Hemorrhagic Septicaemia

Caused by: Aeromonas hydrophila, Pseudomonas fluorescents, etc.

Symptoms: Shallow ulcerations, hemorrhages and in severe cases, swollen abdomen. Internally the body cavity is filled with opaque fluid. Pale liver and sometimes haemorrhages over swim bladder.

Treatment: Water change and check of feed quality. Terramycin (oxytetracycline) with feed 7.5 g/100 kg body weight/day for 10–12 days. Furazolidone 5–7.5 g/100 kg body weight/day for 2–3 weeks. Pond treatment with 3–5 ppm of potassium permanganate.

Fungal Disease

Saprolegniasis

Caused by: Saprolegnia spp.

Symptoms: Ulceration on the skin, fin erosion, exposure of muscles and jaw bones and in some cases tufts of minute white hair like out growths may occur in the affected parts. The disease develops mainly in winter period.

Treatment: Dip treatment in 3 per cent common salt solution for 20–30 minutes or bath in 0.1 ppm Malachite Green.

Diseases Due to Malnutrition and Environmental Stress

Symptoms: Pop-eyes, soft skull and sometimes deformed caudal fins are present., In a later stage of the disease a gradual destruction of the arborescent organs occurs. Knocking on the skull of affected fish produces hallow sound. Delayed calcification and finally breaking of skull. The disease is particularly prevalent in magur larger than 10 cm.

Treatment: Bad pond water conditions such as polluted water, and bad quality feed. The pond water should be exchanged and the flow rate of the water should be increased when the first symptoms of the disease appear. The supply of feed should be stopped for a few days and preferably be replaced by fresh and vitamin C added feed.

Harvesting and Marketing

Partial harvest of magur may start after 50–60 days of culture (except in carp polyculture), when some specimens reach marketable size (above 200 g).

Magur is a clever, fast moving fish. Proper nets are needed for catching magur effectively from the pond. A lot of magur usually escapes through under the bottom line of a net.

Harvesting should be done gently and quickly by seine net preferably in morning when the temperature is cooler. During harvesting, marketable fish should be sorted out first and then small size fish should be released back to pond. In polyculture or magur dominated polyculture the carp species should be sorted out first. The total operation should be done as quickly as possible so that the fishes returned to pond are not stressed.

Harvesting and marketing of fish in rural areas has to be adjusted in accordance with the market days - usually twice a week. Marketing of small quantity of fishes in batches would ensure better price in local rural markets.

References

Bruton, M.N., 1976. On the size reached by *Clarias gariepinus*. *J. Limnol. Soc. S. Africa*, 2: 57–58.

Haniffa, M.A., Jesu Arockia Raj, A. and Arul Mozhi Varma, T., 2001. Optimum rearing conditions for successful artificial propagation of catfish. NBFGR–NATP Publication No.3. Captive breeding of aquaculture and fish germplasm conservation. Paper No. 4.

Hogendoorn, H., 1979. Controlled propagation of African catfish *Clarias tazera* (C and V). I. Reproductive biology and field experiment. *Aquaculture*, 17(4): 323–333.

Islam, M.N., Rahman, S.M., Hossain, Q.Z., Ahsan, M.N. and Asaduzzaman, S.M., 2004. Effect of live and formulated diets on growth and survival of *Clarias batrachus* larvae. In: *Proc. 14th Biennial Nat. Conf.*, The Zoological Society of Bangladesh, Dhaka, 26–27, February, pp. 8.

Janseen, J., 1987. Hatchery management of the African Clariid Catfish *Clarias gariepinus* (Burechell, 1822). In: *Selected Aspects of Warmwater Fish Culture*, (Eds.) A. Coche and D. Edwards. FAO/UNDP, Rome, 1989, 181 pp.

Joseph E. Morris. *Best Management Practices for Channel Catfish Culture*. Department of Animal Ecology 124 Science II, Iowa State UniversityAmes, IA 50011–3221.

Micha, J.C., 1973. Eludedis populations piscicoles de I. vbangui et tentatives de selection et d' adaptators de quelques especes a l'etany de piscicultue-Nogent-san-Marine, Centre Technique Forestier Tropical, 110p.

Morris, J.E., 1993. *Pond Culture of Channel Catfish in the North Central Region*. North-Central Regional Aquaculture Centre.

Munshi, J.S.D., 1996. *Ecology of Heteropneustes fossilis: An Air-breathing Catfish of South-east Asia*. Narendra Publishing House, Delhi, pp. 174.

NACA, 1989. *Integrated Fish Farming in China*. NACA Technical Manual 7. A World Food Day Publication of the Network of Aquaculture Centres in Asia and the Pacific, Bangkok, Thailand. 278 pp.

New, M.B., 1987. *Feed and Feeding of Fish and Shrimp*. ADCP/REP/87/27, FAO/UNDP, 275pp.

Quazizahangir Hossain, M. Altaf Hossain and Parween, Selina, 2006. Artificial breeding and nursery practices of *Clarias batrachus* (Linnaeus,1758). *Scientific World*, 4(4): 82–87.

Raghavan, Rajeev, 2006. Potential culture species in southwestern India. *Global Aquaculture Advocate*, pp. 68–69.

Rao, G.R., Tripathi, S.D. and Sahu, A.K., 1994. *Breeding and Seed Production of the Asian Catfish Clarias batrachus (Lin.)*. Central Institute of Freshwater Aquaculture, Barrackpore, pp. 47.

Saha, M.R., 1996. Effects of various doses of ovaprim for breeding of *Clarias* spp. in Tripura. *Journal of the Inland Fisheries Society of India*, 28(2): 75–84.

Saha, M.R., Mollah, M.F.A. and Roy, P.K., 1998. Rearing of *Clarias batrachus* (Linn.) larvae with formulated diets. *Bangladesh Journal of Fisheries Research*, 2(1): 41–46.

Saha, M.R., Mollah, M.F.A. and Roy, P.K., 1998. Growth and Survival of Clarias batrachus (Linn.) Larvae fed on Formulated Diets. *Bangladesh Journal of Fisheries Research*, 2(2): 151–158.

Sahoo, S.K., Giri, S.S., Chandra, S. and Sahu, A.K., 2007. Spawning performance and egg quality of Asian catfish *Clarias batrachus* (Linn.) at various doses of human chorionic gonadotropin (HCG) injection and latency periods during spawning induction. *Aquaculture*, 266(1–4): 289–292.

Sambhu, C., 2004. African catfish, *Clarias gariepinus* (Burchell, 1822): An ideal candidate for biowaste management. *Indian J. Exp. Biol.*, 42(12): 1226–1229.

Singh, A.K. and Mishra, A., 2001. Environmental issue of exotic catfish culture in Uttar Pradesh. *J. Environ. Biol.*, 22(3): 205–208.

Tripathi, Satyendra Datt, 1996. Present status of breeding and culture of catfishes in South Asia. *Aquat. Living Resour.*, 9: 219–228.

Vijayakumar, C., Sridhar, S. and Haniffa, M.A., 1998. Low cost breeding and hatching techniques for the catfish *Heteropnuestes fossilius* for small-scale farmers. *NAGA*, 21(4): 15–17.

Viveen, W.J.A.R., Richter, C.J.J., Van Oordt, P.G.W.J., Janssen, J.A.L. and Huisman, E.A., 1986. *Practical Manual for the Culture of the African Catfish (Clarias gariepinus)*. University of Wageningen, Netherlands, 121 p.

Yasmin, A., Mollah, M.F.A. and Haylor, G.S., 1998. Rearing of Catfish (*Clarias batrachus*, Lin.) Larvae with Live and Prepared Feeds. *Bangladesh Journal of Fisheries Research*, 2(2): 145–150.

Chapter 19

Cage Culture: Future Potential Culture System

☆ *Mangesh Gawde, Abhijeet Thakare and Gauri Sawant*

Rearing of fin fishes and shell fishes in cages in open water bodies gaining lots of attention through out the world or intensive exploitation of untapped inland and marine water resources. These confined net enclosures used for growing fry/fingerlings of fishes to marketable size, which allows free circulation of water and minimise efforts of maintaining water quality as in pond ecosystem. The dissemination of the cage culture technology needs serious attention from research and extension agencies fisheries sector as well as State Government to increase the current fish production levels through popularization of cage systems to increase standard of living of common farmer and fisherman.

India is gifted with nearly 2.25 million ha ponds and tanks, 3.15 million ha reservoirs with production and beels, swaps and other derelict water bodies of 0.78 million ha which are under exploited and need to harness for augmenting fish production which are producing merely 29.70 kg/ha which is very low as compare to China (743 kg/ha), Sri Lanka (300 kg/ha), Indonesia (177 kg/ha), Cuba (100 kg/ha) USSR (88 kg/ha) and Thailand (64.5 kg/ha). Cage culture is a best alternative where possibility of recapture of reared fishes is very low especially in medium or large reservoirs and larger ponds. Cage culture can be practiced on varied scale depending on available resources as extensive, semi-intensive and intensive scales. The major advantages of cage culture system is its benefits with possibility of making maximum use of available resources with a rapid, easy and sure harvest of culture species and elimination of risk of loss of fishes due to predatory animals.

Site Selection

Site selection is extremely important for success of cages. Rocky substrate is good since it indicates good current scour with reduced risk of waste build up especially for floating cages which reduce the risk of increase of bacterial load within culture system because of accumulation of unused feed.

Various topographical, biological and chemical parameters needed to be studied before selecting site for cage culture and according to that type of cage stationary or floating can be decided.

Some criteria before selection of site are given in Table 19.1.

Table 19.1: Criteria for Site Selection of Cage Culture

Sl.No.	Parameters	Acceptable Standards
A)	**Topographical criteria**	
1.	Height of wave–Stationary Cages	< 0.5 m
	Floating Cages	< 1.0 m
2.	Wind Velocity–Stationary Cages	< 5 knots
	Floating Cages	< 10knots
3.	Water current–Stationary Cages	< 100 cm/sec
	Floating Cages < 100 cm/sec	
4.	Depth–Stationary Cages	4-6 m
	Floating Cages	Minimum 5 m
B)	**Physical criteria**	
1.	Current velocity	10-100 cm/sec
2.	Suspended solids	>10 mg/l
3.	Water temperature	27-31 °C
C)	**Chemical criteria**	
1.	Dissolved Oxygen	> 4 ppm
2.	Ammonia nitrogen	> 0.5 ppm
3.	pH	6.8 - 8.2
4.	Nitrate (NO_3-N)	< 200mg/l
5.	Nitrate (NO_2-N)	< 4 mg/l
6.	Phosphate	< 70mg/l
7.	Chemical Oxygen Demand (COD)	< 3mg/l
8.	Biological Oxygen Demand (BOD)	< 5mg/l

Selection of Species

Selection of species is important in cage culture such as fast growth rate, high Food Conversion Ratio (FCR), high survival rate, quality flesh, resistance to disease and bacterial infections and rapid adaptation to artificial feed. Indian Major Carps, Chinese Carps and Air breathing catfishes are the important species to look for cage culture in fresh water bodies whereas Asian sea bass (*Lates calcarifer*), mullets (*Liza spp.*) and groupers (*Epinephelus malabaricus*) could be acclimatized and adopted for cage culture in brackish and marine water bodies. Disease free seed should be used for stocking. Stocking density varies with primary productivity of water and type of culture practice going to employ. Stocking density of 6-8 fingerlings/m^2 in extensive culture, 12-16 fingerling/m^2 in semi-intensive culture and 25-30 fingerlings/m^2 in intensive culture practice. Periodic checking of stock to observe infected fishes for treatment, removal of dead fishes is necessary to avoid heavy mortality. Regular length-weight sampling is important feed management point of view. Feeding behaviour of fishes is the best indicator to observe health of fishes. Frequent cleaning and wear-tear of cage net bag is vital to minimise static force and ensuring possibility of escape of fishes. Feed should be given @ 20 to 4 per cent and 80 to 10 per cent of body weight if artificial feed is given and raw fishes in case of Asian sea bass respectively.

Cage Design and Construction

Shape of cage doesn't make significant difference in fish production, its important that cage volume remain relatively resistant to deformation by external forces that could cause crowding, stress and mortality of stock. Static force is vertical which includes weight of net bag and fouling. Static force has to enumerate for design of flotation and mooring system. The size of fish cages depends on the scale of operation, species and financial and managerial resources. Cage depths between 1 to 2 m are generally considered ideal for most fresh water fishes whereas 2-3 m for marine fishes, which gives stock sufficient shelter from surface effects and providing adequate volume for exercise and feeding. Mesh of cage net depends upon size of fishes planned to stock, 4-6 mm and 16-20 mm mesh webbings is most commonly used for grow-out stages. Rectangular cages of 50-m^2 area or circular cages of 10 m diameter can be use for commercial rearing of fishes.

Cage Design

Cage design depends upon various factors like species to be cultured prevailing environmental conditions, method of culture (extensive/semi-intensive/intensive), costs and availability of material as well as local skills.

Two main components of cage design are:

1. Cage structure
2. Cage size

Cage Structure

Cages are composed of a frame, netting or floats, sinkers and anchors. The frame supports the cage netting, maintains the shape of the cage, and acts as a float or operational platform. Traditional cages are fabricated using existing skills from locally available material like grasses or wood. Bamboo is much more suitable material being strong, cheap, widely available material like grasses or wood. Bamboo is a much more suitable material being strong, cheap, widely available and easy to use. But these materials decompose at a faster rate in tropical water. Caged fish and predators can also damage them. Modern cages utilize materials such as nylon, plastic polyethylene and steel mesh that are though more expensive, have much longer life span and permit better water exchange.

Cage Size

Cage size is ultimately based on the trade-off between cost of fabrication, operational cost and per unit area of production, and efficiency of production. It depends on various factors- the larger the cage areas lower the construction cost. Here, the area of net material required is 30 per cent higher when small cages are constructed rather than one large cage with same total surface area. The productions of larger cages are relatively lower than small cages because of inadequate scope for utilization of natural food.

Mesh Size

Cage mesh size should be selected in such a way that stocked fish cannot escape, and a better water exchange is ensured. Environmental criteria for cages: -Sites with violent storms and extreme cold should be avoided. Cage culture has immense potential to create employment opportunities to youth of rural areas with keen attentions to awareness campaigns and training on technical know how's. Cage culture can be the best alternative to achieve targeted per capita consumption of fishes from all available resources that cannot be utilised for culture or culture-based-capture fisheries. The private sector can enter in cage culture system by using large size cages or linkage of small cage structures for meeting mounting demand of fishes in urban areas. With extensive cage culture practice we can get production of 6 kg/m^2 and 15 kg/m^2 and 28 kg/m^2 from semi-intensive and intensive cage culture practices respectively with 10 months of culture period.

As *Lates calcarifer* is being highly carnivorous fish and cannibalistic in nature. In order to reduce losses due to cannibalism, grow out is performed in two phases the fingerlings are grown to a weight of 20g in special nursery cages. Besides the natural food, fishes are fed with supplementary feed consisting of trash fish twice a day. The rearing period is 30-45 days. Periodically, the fishes of similar size are removed and stocked in separate grow-out cages for growing to marketable size. We can stocked @ 25-30 no/m^2. After three months - tinned out to 10-20 fish /m^2. The artificial feed is given @ 15 per cent of the body weight initially and reduced to 5 per cent of body weight for next 3 months. When insufficient trash fish is not available; rice bran and broken rice is added as a partial substitute.

Thus, cage culture of fishes holds a great promise as future potential system for large-scale fish production from untapped resources and employment generation.

Chapter 20

Crafts and Gears Used in Yeldari Reservoir, Maharashtra

☆ *S.D. Niture and S.P. Chavan*

Introduction

The Yeldari reservoir is a large sized reservoir of about 6,272 ha. area, constructed on Purna river in Jintur tahasil of Parbhani district of Maharashtra. Reservoir area is included in the survey of India toposheet map No. 56A/10. It was the first ever major project in Marathwada region for irrigation and power generation to initiate the process of economic development of Marathwada region. It is mainly hydro-electric project; therefore all rights of reservoir management and maintenance are towards the State Irrigation Department and State Electricity Board of Maharashtra State.

Gill net, Cast net, Extra large Cast net, Drag-Bag net, Hand net, Hooks and line, are used to catch the fish and prawns. To operate the fishing nets, the thermocoel rafts of 1 to 3 person carrying capacity are used. The use of mechanized boats in the fishing was only to carry the harvested catch from the various fishing sites around the reservoir to main collection centre.

Materials and Methods

Important landing centres of reservoir were visited regularly and information on methods of fishing in reservoir was collected. Beside this, a questionnaire method was adopted to collect data on size of nets, its types and operation. Information was also collected on different types of crafts used in fishing.

Results and Discussion

The details of the gears and crafts used for fishing in Yeldari reservoir are described below:

Gears

Gill Net

Gill nets are passively operating nets, once they are set, then these nets remain stationary. It is called gill net because the fish gets entangled in to the mesh in the opercular region.

Gill nets are constructed with floats made up of from soft wood, plastic pieces or thermocoel pieces fixed at regular intervals to the upper line of the net where as stone pieces or metal pieces are used as sinkers, which are attached at regular intervals in the lower line of the net.The mesh size varies from 1.4 cm. to 16 cm. The gill net is made up of from cotton or transparent synthetic fibers, which are commonly called as 'Nylon net' or 'Disco net'. Depending upon variation in mesh size, it is commonly called as 'Ek boti jal', 'Do boti jal', 'Char boti jal', 'Panch boti jal', 'saha boti jal' etc.

Readymade webbled gill nets were purchased from Yeldari-camp market for the cost of Rs. 400-500/kg. Do boti jal', 'Teen boti jal', 'Char boti jal', and 'Panch boti jal' readymade webbled gill nets were for the cost of Rs. 500 /kg; where as 'Saha boti jal', 'Sath boti jal', 'Aath boti', 'Nau boti', 'Daha boti', and 'Bara boti' ready made gill nets were for the cost of Rs. 400 /kg.

The gill nets are of various dimensions *i.e.*, 20 m to 100 m in length and 2 m to 15 m in height. The quality of Nylon fiber used for the preparation of these types of net is very poor, therefore the fishermen change their gill nets within 3 to 4 months due to tearing of nets.

The gill nets are generally payed off in the evening and generally hauled up next morning. Every fisherman arrange about 6-15 kg. gill net every day at different places in the Yeldari reservoir. Some times 2 to 3 fishermen arrange their gill nets by uniting the gill net pieces up to 400 - 500 meter length by using 3 to 4 thermocol rafts and bamboo sticks flippers to propel the rafts. The long bamboo are also used to propel the rafts by pressing one end of the bamboo against the bottom of the reservoir. Regularly in the morning, the nets are hauled and trapped fishes and prawns are collected in the mosquito net bags, and once again the nets are spread for fishing into the water and in the evening the nets are hauled to collect the catch.

The fish catch obtain was not definite by using gill net. Some time even a single fish do not found trapped where as in some operations 0.5 to 15 kg fish catch was obtained.Generally after 8 to 10 days the fishermen dry their gill nets around the reservoir shore by sun-drying process.

Cast Net

It is one of the commonly used net by the fishermen of Yeldari reservoir, it is commonly called as 'Phekjal','Ghagra jal', this is circular-mouthed or umbrella-shaped conical net made of cotton- twine webbing of about 2.5 cm mesh size. A strong cord or rope was attached at the apex of the cone of the net. Number of iron made cylindrical nuts were fixed as sinkers all along the circular periphery. Circular pockets were made at the periphery by folding the net inwards to about 6 to 8 meshes in depth. The size of the circular mouth circumference of the net varies from 4 m. to 15 m. To prepare a cast net of medium size the material cost is Rs. 1800 - 2000 and time required to construct the net is about 15 to 20 days for a single fishermen. The traditional fishermen construct this type of net, it is a hand made net and not available in Yeldari camp or other surrounding market places like Jintur, Hingoli, Parbhani, Nanded. The fishermen web this type of net by purchasing the raw materials from the market at Yeldari camp.

During fishing operation the portion near the apex rope of the net is held in the left hand and the middle portion is kept loose and the lowermost portion *i.e.*, the margin of net mouth is held with the other hand. With a swing, the net is lifted up just above the head and thrown with a circular swing in

Figure 20.1: A Caste Net Operated in Yeldari Reservoir

the reservoir to a certain distance according to the length of attached rope to the net. This type of net was used to catch all types of fishes and prawns from the reservoir.

This net is operated by single fisherman and the catch obtained by using cast net is good as compared to gill net. Single fisherman can carry out 10 to 15 operations within a day by using this type of net and the catch obtained by this net varies from 0-30 kg in a single operation.

The fishing attempts in a day for this net has great variation because, if in a fishing attempt when there is a good catch then the post harvest processes are lengthy, when in a fishing attempt if very little catch or no catch then the fishing attempts may increase in number. At the end of fishing day at evening net is cleaned and dried.

Drag-bag Net (Seines)

The Drag bag net used in Yeldari reservoir are commonly called as 'Pandya' or 'Purai' and 'Wadap' or 'Zorli'.These are generally large sized nets of variable dimensions and mesh sizes, as per requirements *i.e.*, water level in the reservoir, availability of budget and man power etc.. In this type of net two lateral flanks of net are connected to the central large elongated quadrangular bag which tapers towards the end. The lateral flanks of the net are rectangular in shape and made up of from 6 to 20 pieces of net, each piece of the net is 2-6 m. in height and 3-7m. in width. The good quality plastic float are fixed to the upper line of flanks of the net while the stone pieces or iron pieces are fixed to lower line of the flanks of the net. There is no use of floats and sinkers to the central bag of this net.

In 'Purai' or 'Pandya' net each lateral flank is made up of from 3 to 10 pieces of net. The mesh size of the net is usually 3 x 3 cm. or varies, it is larger at the outer extremity and smaller towards the mid region. The length of central bag varies from 10-50 m. and the dimension of mouth of the bag varies from 2 x 2 m. to 2.5 x 2.5 m. The maximum height of the entire net as a whole net piece is at the middle region *i.e.*, at the mouth of the central bag, which tapers towards the two lateral flanks. The height of lateral flanks of Purai net was 6 -12 m. and length of two flanks together ranges up to 200 to 500 m.The weight of Purai net was 70 to 150 kg and cost varies from Rs. 35,000 to Rs.1,00,000.

The Pandya net was slightly larger than Purai net.The Drag bag nets like the 'Zorli' or 'Wadap' are also similar in structure like Purai and pandya. the main difference between 'Zorli', 'Wadap' net and 'Purai', 'Pandya' net is that, the Zorli, Wadap nets are made up of mosquito net, hence weight of these nets is less than Pandya and Purai nets. Cost of Zorli or Wadap nets is also less than Pandya and Purai nets and it varies from Rs. 30,000 - 70,000, depending on variation in size of the net.To operate the drag bag net there was variation in the requirement of fishermen number, depending on the variation on the size of the net.4 to 10 fishermen are required to operate these nets.

The terminal of both flanks of the net are either fixed to iron anchor on the coast of reservoir or held in the hands of a fishermen group on the coast of reservoir, rest of the part of the net is lodged on the combined 3 - 4 jointed thermocoel rafts. 2 to 5 fishermen carry the net away from the coast of reservoir and release the net in the water with the help of bamboo sticks. The net is released in semicircular manner so that another flank of the net is carried to the coast of the reservoir. The operational diagram of the net is as shown in the fig. The distance between the terminal points of flanks of the net on coast of the reservoir after encircling the water body varies from 50 to 100 feet. Finally both the terminal ropes of net flanks are dragged by two separate groups of the fishermen or one end remains fixed to the anchor and other terminal rope of flank is dragged.

When the net flanks are dragged towards the coast line of reservoir then the fish and prawns are directed towards central bag of the net.The lower lines of flanks and the lower portion of the mouth of central bag remain at the bottom of reservoir due to presence of weight. The catch enters and get trapped in to the central bag which is removed on the coast of reservoir.

The operation of these nets is similar to the bag- net operation in marine fisheries for sardines, mackerels and bombay-duck catch. Two to three fishing operations are possible by this net in a day, the catch obtained by this net in Yeldari reservoir at various fishing stations varies from 10 kg. to 11 qt. in a day for single net. The catch includes all type of fishes containing weed fishes, predatory fishes, Indian major carps and prawns of variable size.

Hand Lift Net

It was very interesting type of net found to be used in riverine and reservoir fisheries of this region. In the Yeldari reservoir in sandy or muddy, shallow coastal region usually it is operated by the fisherwomen to catch the prawns. The net consist of triangular frame of bamboo of height 5 to 6 ft. There is a small piece of bamboo fixed in the anterior region of the triangular frame of the net, which is used to hold the net and to drag the net. At the front part of the bamboo frame a net piece of 3 to 4 cm mesh size is fixed from the basal bamboo up to small piece of bamboo fixed in the anterior region of the triangular frame, and at the centre a conical mosquito net bag of 8 to 15 feet length is fixed. There is a rope fixed at the two ends of the bottom bamboo of the net frame, which forms a belt like structure, which is fixed around the waist of the fisherwomen. The catch remained trapped in the central bag. The efficiency of the net is catch of 1 - 2 kg. prawns (*Macrobrachium malcolmsonii*) per day.

This is single person operated net. The modified version of this net of quadrangular type may require three persons to operate. The net is dragged on the bottom of water body by keeping the bamboo frame vertical. It is dragged for 50 to 70 meter distance and finally the frame is lifted to collect the trapped catch. Finally the catch collected in central bag is removed by lifting the whole frame of the netfor about 1-2 ft. from the water surface. Partially it is drag net, bag net and lift net. It is named locally as 'Pilna'. The efficiency of the net various from 1-10 kg. prawn catch per day, depending on population density of prawn in the reservoir.

Hook and Line

This type of fishing gear is used to catch the predatory fish species from the reservoir. This equipment contains a long nylon wire or rope of 100 to 150 feet length having knots at 1 to 1.5 feet distanc all along the rope. In between two knots a nylon wire of 3 to 4 feet length is loosely fixed so as to move this wire in between two knots. At the free terminal of the wires rounded, metallic hard hooks are fixed. To each hook freshly caught small sized weed fish species like *Chela sp, Puntius ticto, Amblypharyngodon mola, Ambassis sp.* etc are fixed as bait to attract the predatory fishes.

To the main long rope or line, the plastic or thermocoel floats are fixed. At few intervals the weight in the form of stone pieces are also fixed to the main line. Depending upon the depth of water. there is variation in the weight fixed to the line. It was found that only weed fish species were used as bait but any other type of bait like insects, cockroaches, grubs, earthworms, cattle beef, cattle liver were not in use for this purpose. About 100 - 110 hooks are fixed to a single line.

Usually old fishermen or children are involved to use the hooks and line for fishing. All the hooks are baited with freshly caught (not dry) weed fishes. One terminal of main line is fixed at the shore of reservoir to a large stone or anchor or a shrub and entire line with baited hooks is carried in a bamboo basket and fishermen use thermocoel rafts to carry the line and release the line and hooks into the reservoir water away from coast of the reservoir. Usually the line is released in a straight line or some times in a random direction in the reservoir water.

The predatory fish like *Wallago attu, Heteropneustes fossilis, Mystus seenghala, Mystus bleekeri* usually attracted towards the baited fish, they engulf the bait along with hook and get trapped or hooked.

The operation of line and hook gear depends on availability of weed fishes used as bait, hence there is no fixed time to release the hook and line but after it's release the entire hook and system is removed after every 2 to 3 hours to check the trapped fishes and again released in the same direction or at different direction. The line hooks are checked for increase in weight regularly.

Fishing operation by hooks and line are possible for 3-4 attempts in a day. The efficiency of this gear depends on population of predatory fishes. This is an effective method for the eradication of predatory fishes from the reservoir too.

Crafts

Thermocoel Raft

The craft used in Yeldari reservoir is locally called as 'Nav'. It is made up of from thermocoel sheet of size 7 x 2.5 ft. long with 6 to 8 inches thickness. The Nav is prepared by using single continued sheet of thermocol or two thermocoel pieces of size 3 x 2.5 ft. Each sheet is covered or packed tightly by mosquito net.The cost of one craft is about Rs. 450 - 550 in the market at Yeldari camp. During 2005 to 2008 study duration almost all crafts used were of this type. The fishermen purchase the crafts individually.As these crafts are not suitable to carry large sized heavy nets like Pandya, Purai, Zorli, or Wadap, hence 3 to 8 thermocoel rafts were connected to each other from their laterally to form large sized platform. The thermocoel raft is navigated by single fisherman with the help of a hand flipper which is made up of from single bamboo stick of 5 feet, to which the plastic flippers are fitted at both the end of bamboo. The flippers are prepared from cutting the PVC (Poly Vinyl Chloride) pipe pieces, the dimensions of the flipper blade are 1 x 0.5 ft. The flipper is locally called as 'Chatu'.

To drive this boat with the help of flipper need a good balance technique and practice. This craft found to carry 30 to 40 kg stone pieces to tie with the nets as sinkers, it also carry a net of 4-5 kg weight. The catch obtained in gill net is collected in a nylon happa bag or mosquito bag. This type of bag is tied to the raft from posterior side and the bag containing catch of 30-40 kg. carried by dragging the bag in water, but not carried on the raft platform. It is single fisherman carrying raft.After fishing the rafts are sun- dried on the coast of reservoir

Masula Type Metal Sheet Non-mechanized Boat

This type of boat was also found to be used by the fishermen in fishing in the Yeldari reservoir. It is constructed from Iron metal sheets fixed with welding and rivets. In these boats at some places wooden planks were also used. The length of this craft was 10-15 feet and width at the centre was 5 to 6 feet. The base of this craft was flat.The carrying capacity of this crafts was 4 to 5 person. The weight of this craft was 80 to 100 kg. This craft is operated by using one or two flipper by 1 to 2 fishermen.These crafts were not so famous due to high construction cost of Rs.8,000–10,000 rupees. These types of craft were 1 to 2 in number only.

Mechanized Boat

Petrol Engine Speed Boat

This is a fiber-plastic small sized mechanized boat. There is 350 c.c. petrol engine fixed to this boat. The boat was having a steering wheel and posterior flipper like manually operated structure to change the direction of the movement of the boat. The speed of this boat was 10 to 25 nautical miles/ hr.

It carries 8 to 10 person. In the Yeldari reservoir this type of boat was firstly used by Purna society to control the fish poaching and control on type of fishing nets used by the fishermen. It was a vigilance squad boat. It was one in number.

Diesel Engine Boat

It was metallic large sized mechanized boat of total length 20 feet and height 7 feet, width at the centre was 10 feet. It was modern type of boat fitted with a diesel engine of similar to the capacity of truck engine; it was having a chamber to store the catch of capacity 10 to 15 qt. It carries 25 to 30 fishermen. It was 1 in number at Yeldari reservoir. It was planned to use in fishing activities by the Purna Society but it was found to be used very rarely.

Purchase and Maintenance of Nets and Boats

The nets used at Yeldari reservoir by the individual fishermen and group of fishermen are purchased from the net markets at Nanded, Mumbai and Hyderabad. The nylon yarn, sinkers, floats and disco nets of readymade mesh type and other accessories for the net construction are purchased

Figure 20.2: An Engine Boat Operated in Yeldari Reservoir

by individual fishermen from small shops of nets at Yeldari camp. It was observed that, three net shops were present at Yeldari camp.The thermocoel sheets and plastic mosquito net covers to construct the thermocol rafts were also purchased by individual fishermen from the shops of Yeldari camp. After fishing the torn mesh and broken lines of net and damaged floats are repaired by the fishermen. After fishing the nets and thermocol rafts are sun dried. The nets are packed in nylon mosquito-net bags and stored. Except sun drying there was no any processing for the maintenance and preservation of nets and crafts.

Chapter 21

Socio-economic Status of Fishermen Community Around Bori Tank Near Naldurg, Maharashtra

☆ *M.G. Babare and M.V. Mote*

Introduction

In fisheries, social means a relation to the interaction of human beings with each other, as individuals or as groups. Economics is one form of interaction between people and regarded on one of the social science. Socio-economic is common combination encountered in the field of planning and development work. Socio-economic studies begin with a detailed description of the socio demographic characters of population.

In the fisheries sector surveys had been conducted by various agencies and research workers in different regions of our country to study one or the other problems of the fishermen community.

The notable workers are Desai and Baichwal (1960) Choudhary (1989), Rout and Das (1992), Sathidas and Panikkar (1988), Sakhare (2003), Venkat Raman (1981). These workers focused on the socio economics status of fishermen around different reservoirs in India.

There is no any back record about the socio economic status of fishermen around the Bori tank, Naldurg hence this investigation was under taken.

Study Area

The Bori tank was constructed in the year 1967 situated near the Naldurg town, Osmanabad district (M.S.) India. The tank area is bounded by latitude 70°–16′N and longitude 70°–50′E. The catchments area of Bori tank is 9.39 km. The mean depth of the tank is 18.46 meter indicates shallow and deep characters of pond. The water of the tank is used for Agriculture, drinking and Fish–culture purposes.

Material and Methods

The Bori tank is an important water resource of this region. Fishing is carried out throughout year. The present communication deals with the Socio-economic status of fishermen around Bori tank. A questionnaire method was adopted to collect data on socio-economic status of fishermen community. Fish catch per day and monthly income earned by fisherman for this personal contact was used. Annual audit report and district fishery development office was provides economical statements. The survey was conducted during the year 2008.

Results and Discussion

The following parameters were used to study the socio-economic status of fishermen around the Bori-tank.

Family Size and Caste

The size of the families are slightly big one. Average size is 4 to 7 people in one family. The castes of the fishermen are Bhoi, Banjara, and some other backward classes were regularly practiced fishing. The languages of the Banjara people are Banjara (lamani) and they can speak and write in Marathi languages Bhoi people can speak and write in Marathi.

Housing

On the bank of the tank a village named Vasant Nagar had named after the name of late C.M. of Maharashtra Vasantrao Naik. Most of the people are living in pucca houses. A round 62 per cent people are living in Pucca houses where as 31 per cent. People are living in Kutcha houses and remaining 7 per cent people use huts.

Educational Status

In the present study area, one residential high school one primary school is working. With regard to educational level, it could be observed that 75 per cent per cent the respondents were literate while only 25 per cent are illiterate. Among the literate schooling up to high school level occupied 45 per cent followed by middle school level, pre degree level, primary level and graduate level with percentage of 19, 16, 13 and 8 respectively. Educational status affects much more in fish farming practice.

Occupation

As the standard of living and earnings of people depends upon their occupation. On an average 70 per cent people are engaged in agriculture while 20 per cent, 6 per cent and 4 per cent people are engaged in fishery sector, business and service respectively.

Age

Age is another important factor of socio-economic of a community. In the present study area 47 per cent of the people belongs to middle age group followed by 38 per cent younger group and 8.33 per cent older age group. Younger and middle age group are more interested in fish farming than old age group.

Experience

Experience in any field is age dependent factor.' Young people have less experience than the middle age group while middle age group had less experience than old age group. Always young and middle age group was working in fisheries under the guidance of old age people who can guide than well.

Social Participation

The fishermen community of Vasant Nagar Village participated in social institution like co-operative, village welfare organization, school, library, group festivals etc. About 75 per cent of the people shown active participation in social programs.

Training

Training is an effective tool for transfer of technology even through training programs are being organized by the FFDA's and other agencies, the fishermen were not willing to under go training due to fear of wage loss, lack of time and lack of incentive. Majority of the fishermen did not receive training on fish culture practices. The number of trained respondent in Vasant Nagar village were 25.

Total Family Income

As " Banjara or Laman" community belongs from the poor community most of the people were below the poverty line. In general, employment and income are the twin decisive factors mostly used for determining the living standard of any community or region. Equitable distribution of income factor enhances the social harmony among different sections of population. Analysis of income level of Vasnat Nagar village studied has brought out some interesting features. On an average fishermen annual income levels are distributed as less than 10000 Rs. 40000-50000 Rs. and above 100000 Rs. People belong to 10000 Rs. 40000-50000 Rs. and above 100000 Rs. are 67 per cent, 21 per cent and 12 per cent respectively. This low level of income reflects their poor economic condition, which was not sufficient to maintain their normal livelihood. They can't afford much for fish culture activities.

Total Family Expenditure

Most of the fishermen were in the low-income trap and they couldn't meet even their consumption expenditure from their earnings. The average annual expenditure of a fishermen household works out at Rs. 27000 per annum. The perusal of expenditure pattern clearly indicates that about 65 per cent of there total income spends for their food alone.

The clothing was found to be next major expenditure among the fishermen the low level of spending for education and medical indicate there socio-economic backwardness.

Source of Information

In the present study selected village identifies the sources of information for fish culture practices as fish seed vendors, friends and relative, fisheries extension persons, radio, TV and newspapers. On an average large majority of the respondent got their information from fish seed vendors followed by extension persons, radio, TV and newspapers.

References

Choudhury, M., 1989.An econometric study on the socio-economic status of the fishermen community in lower Assam. *J. Inland Fish. Soc. of India*, 2(1): 7–13.

Desai, M.B. and Baichawal, P.R., 1960. Economics survey of fishery industry in Thana district, Bombay, India. In: *The Economic Role of Middlemen and Co-operatives in Indo-Pacific Fisheries*, (Ed.) E.F. Szezepanic. FAO, Rome.

Rout, P.K. and Das, S.K., 1992. Socio-economic aspects of extensive shrimp agriculture along Gopal Pur coast in Ganjam district of Orissa. *Seafood Export Journal*, 24(J): 5–8.

Sakhare, V.B., 2003. Socio-economic status of fisherman around Yeldari reservoir, Maharashtra. *Aqua Tech.*, p. 77–78.

Sathidas, R. and Panikkar, K.K., 1988.Socio economics of small scale fishermen with emphasis on costs and earnings of traditional fishing units along Trivendram coast, Kerala: A case study. *Seafood Export Journal*, 20(2): 21–36.

Siddiqui, R., 1996. A study of socio-economic problems of the fishermen in Tamil Nadu and Orissa. *Indian Journal of Extension Education*, 21: 86–88.

Chapter 22

Success Rate of Fishing in Potential Fishing Zone (PFZ) Grounds Over Non-Potential Fishing Zone (Non-PFZ) Grounds

☆ *A.U. Thakare and M.M. Shirdhankar*

Introduction

The Catch Per Unit Effort (CPUE) of all kinds of marine fishing fleet is decreasing day-by-day as the number of fishing vessels of all categories has increased remarkably making the fishing business uneconomical. Amongst the major fishing activity, purse-seining is playing major role in the recent past. Success of this fishing depends mainly upon the experience of fisherman particularly with regard to his senses such as sighting of shoals, odour- and smell- detection produced by varieties of fish shoals, etc. Shoaling fishes such as sardine, mackerel, seer fish, tuna etc. are the main catches of purse-seine.

Potential Fishing Zone (PFZ) forecast is generated by Indian National Centre for Ocean Information Services (INCOIS) stationed at Hyderabad for the entire coast of India. The Potential Fishing Zones are located on the basis of chlorophyll concentration and Sea Surface Temperature (SST). It is expected that if the fishing vessels are directly guided by Potential Fishing Zone forecast, it will increase the success rate of fishing. This in turn will lead to save time and also reduction in the diesel requirement for fishing. In this background, the present study was undertaken along the Ratnagiri coast of Maharashtra to know the extent of *in-situ* validity of PFZ forecast made by the INCOIS.

Methodology

Study area from Shrivardhan to Vengurla along the West coast of India situated in Maharashtra state with 288 km coastline was selected for the present study. Mirkarwada being one of the important fishing harbours in this area was selected as base-port in the study area. Coastal area between 18° 00′ N 72° 00′ E to 15° 43′ N 73° 33′ E was selected for the study. A total of 15 Purse-seiners were selected as sampling units and were coded as MP_1 to MP_15 (Mirkarwada Purse-seiner). The PFZ forecast was given to the sampling units and feedback was collected after each cruise. The data of 15 purse-seiners was collected from October 2006 to April 2007 was collected and was analysed for the fishing success rate of vessels fished in PFZ over the vessels fished in the Non-PFZ area.

Results and Discussion

Fishing success rate of vessels fished in PFZ was calculated to check the reliability of PFZ forecast over the vessels fished in Non-PFZ (Table 22.1). Higher fishing success rate was observed in the month of November (93.33 per cent), whereas lower success rate (46.67 per cent) in the month of April (Figure 22.1). Success rate recorded was always more than 60 per cent for all the months except in the month of April (46.67 per cent) and on an average 72.79 per cent success rate was observed during the complete fishing season.

Table 22.1: Month-wise Success Rate of Purse-Seiners Fished in PFZ

Month	*Success Rate (Per cent)*
October	92.86
November	93.33
December	73.33
January	60.00
February	73.33
March	70.00
April	46.67
Overall	72.79

Fishing success rate was also calculated for individual purse-seiners operated in PFZ over purse-seiners fished in Non-PFZ. Fishing success rate of the purse-seiners varied between 42.86 to 100 per cent with overall fishing success rate of 72.95 per cent (Table 22.2 and Figure 22.2). Higher success rate was shown by purse-seiner MP_5 (100 per cent), whereas lower success rate was recorded for the purse-seiner MP_10 (42.86 per cent). In general, the fishing success rate of more than 50 per cent was recorded by all the purse-seiners except MP_10.

The fishing success rate in any of the month of the purse-seiners fished in PFZ was in the range of 46.67 to 93.33 per cent with an average value of 72.79 per cent, whereas fishing success rate for the individual purse-seiners fished in PFZ was in the range of 42.86 to 100 per cent with an average of 72.95 per cent. Solanki *et al.* (2004) have reported about 85 per cent success rate in the PFZ areas, whereas Dwivedi *et al.* (2005) have reported 70 to 90 per cent success rate for bottom trawling and gill net fishing.

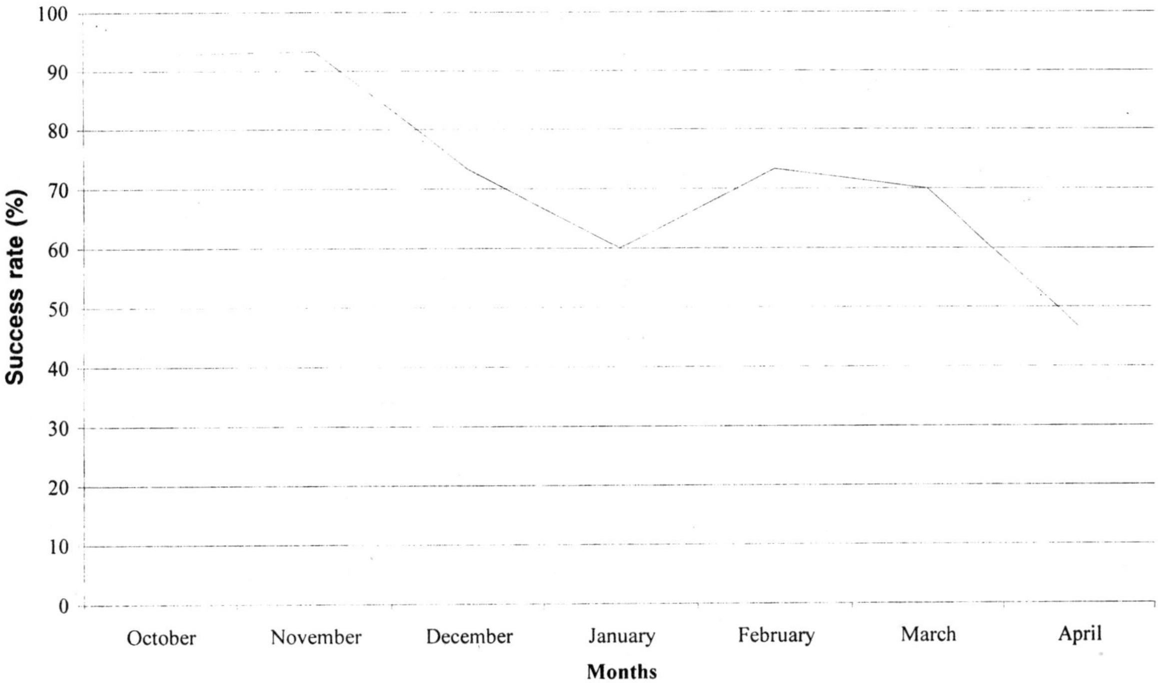

Figure 22.1: Month-wise Percentage Fishing Success rate in PFZ over Non-PFZ

Table 22.2: Vessel-Wise Success Rate in PFZ

Vessel Code	*Success Rate (Per cent)*
MP_1	80.00
MP_2	85.71
MP_3	85.71
MP_4	83.33
MP_5	100.00
MP_6	71.43
MP_7	71.43
MP_8	71.43
MP_9	57.14
MP_10	42.86
MP_11	71.43
MP_12	66.67
MP_13	50.00
MP_14	85.71
MP_15	71.43
Overall	72.95

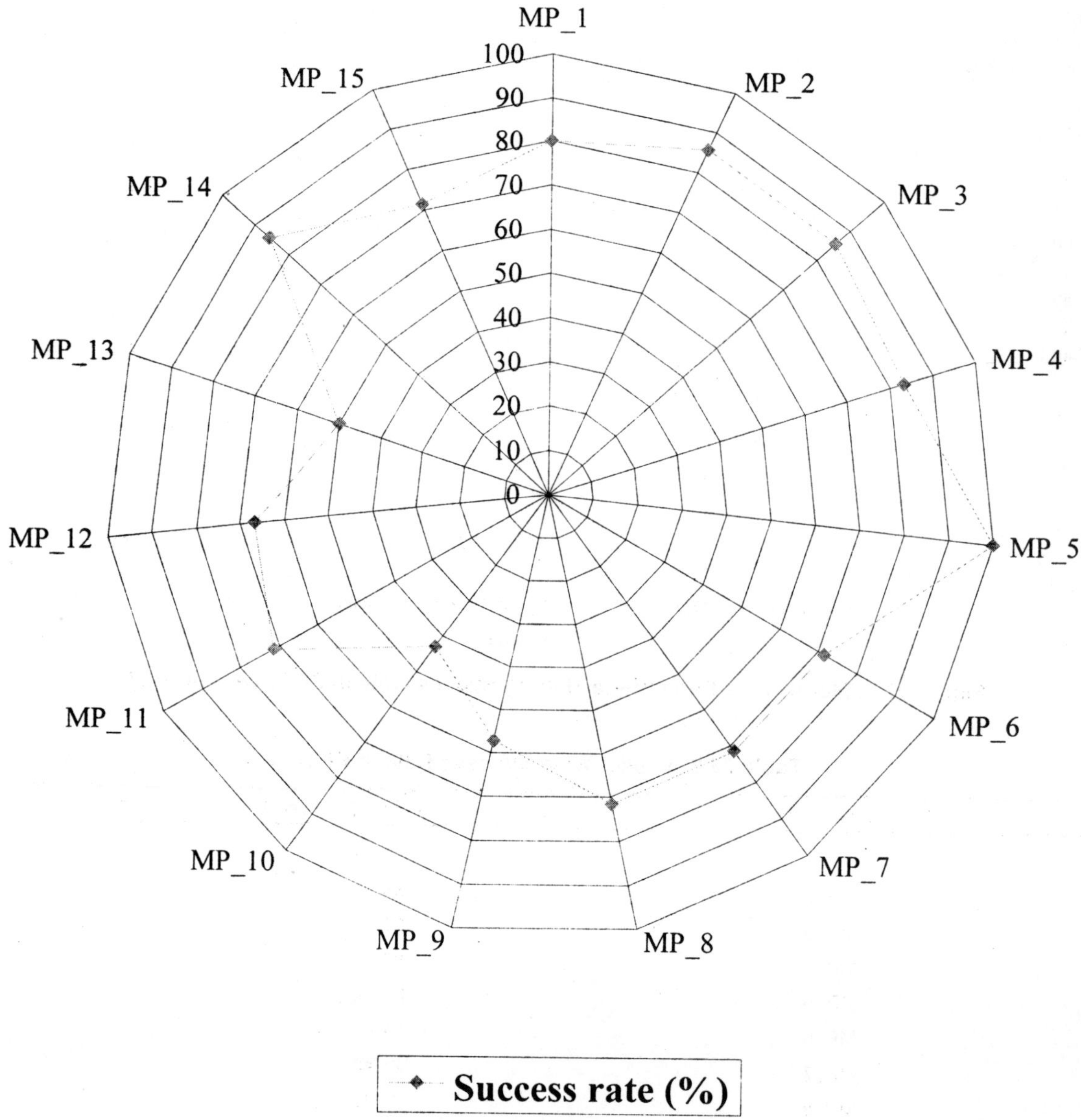

Figure 22.2: Vessel-Wise Percentage Fishing Success Rate in PFZ Over Non-PFZ

Conclusion

The fishing success rate of purse-seiners fished in PFZ recorded during complete fishing season was in the range of 46.67 to 93.33 per cent with seasonal average of 72.79 per cent. Individual purse-seiners fished in PFZ showed fishing success rate varying between 42.80 to 100 per cent with an average value of 72.95 per cent.

From the present study, it can be derived that the catches obtained in PFZ as per the INCOIS forecast were comparatively higher than that of Non-PFZ. This observation is further supported by increased success rate observed in PFZ over Non-PFZ. Thus, the dissemination of information with regard to PFZ would be helpful to fisher population for increased time, man-power and economic efficiency.

References

Ben-Yami, M., 2004. *Purse-Seining Manual*. Fishing News Book, Osney MeadOX2 0EL, England, 406 p.

Dwivedi, R.M., Solanki, H.U., Nayak, S.R., Gulati, D. and Somvanshi, V.S., 2005. Exploration of fishery resources through integration of ocean colour with sea surface temperature: Indian experience. *Indian Journal of Marine Sciences*, 34(4): 430–440.

Snedecor, G.W. and Cochran, W.G., 1967. *Statistical Methods*, 6th Edition. Oxford and IBH Publishing Co., New Delhi, 593 p.

Solanki, H.U., Dwivedi, R.M. and Nayak, S.R., 2004. Application of remotely sensed closely coupled biological and physical processes for marine fishery resources exploitation. *International Journal of Remote Sensing*, p. 1–6.

Chapter 23

Fisheries Management of Yeldari Reservoir, Maharashtra

☆ *S.D. Niture and S.P. Chavan*

The Yeldari reservoir is a large sized reservoir of about 6,272 ha. area, constructed on Purna river at Yeldari camp in Parbhani district of Maharashtra.The reservoir lies in between North Latitude 19°-43′-00″ and East Longitude 76°- 45′- 00″. It is naturally situated in hilly region on its both sides. Reservoir area is included in the survey of India toposheet map No. 56A/10. It was the first ever major project in Marathwada region for irrigation and power generation to initiate the process of economic development of Marathwada region. The catchment area of reservoir is 7,330 km^2. It is mainly hydro-electric project, therefore all rights of reservoir management and maintenance are towards the State Irrigation Department and State Electricity Board of Maharashtra. The reservoir consist of 3 outlets, through which the water is released in to the Purna river basin, after passing through the turbines of electricity generation plant and ten spill-way gates. The discharge of water from the Yeldari reservoir rotate the turbine for hydro-electric power generation; this project is effectively functioning from the year 1965 and it was managed by the Irrigation Department up to the year 1972, but later on it is managed and governed by the Maharashtra State Electricity Board. This project was constructed by using Swedish technology with capacity of 22.50 m.w. electricity generation through 03 electricity generating units, each set of 7.5 m.w. capacity. Whenever water from the reservoir is released for irrigation the hydro-electric units start functioning because the released water is passed through the specially constructed tunnels and passed over the turbines for electricity generation. The principle features of Yeldari reservoir are depicted below:

The Principle Features of Yeldari Reservoir

1. General

I. Name of Dam	Yeldari
II. Year of start	1958

III. Final Completion	Oct 1968
IV. Design purpose	Hydropower and storage
V. Location	
a) Latitude	19°–43′–00″ N
b) Longitude	76°– 45′–00″ E
c) Village	Yeldari.,Tq.Sengaon Dist. Hingoli Yeldari camp, Tq. Jintur Dist. Parbhani
2. Hydrology	
I. River	Purna
II. Sub-basin	Purna
III. Basin	Godawari
IV. Gross Catchment area	7329.7 sq km
V. Average Rainfall	685 to 762 mm
3. Reservoir	
I. Gross storage (F.R.L)	934.44 M cum
II. Gross Storage (M..W.L)	1029.92 M cum
III. Live Storage	809.77 M cum
IV. Dead Storage	124.67 M cum
V. Carry Over	243.52 M cum
VI. Water spread (M.W.L)	106.64 sq km
VII. Water spread (F.R.L)	101.54 sq km
VIII. Annual evaporation	128.2 M cum
IX. Sediment rate M cu/sq km/year	317.5
X. Life of Reservoir	75 years
XI. Maximum fetch in km	25
Dam	
1. Types of Dam	Earthen dam with Masonry Spillway in gorge portion.
2. Top of Dam	465.89 M
3. M.W.L	462.38 M
4. F.R.L	461.77 M
5. Crest level	450.19 M
6. M.D.D.L	447.75 M
7. Irrigation outlet Sill	441.96 M
8. Length of Dam	
a) Total length	4432 M
b) Earthen Dam	4880 M
c) Overflow Section	149.65 M

9.	Types of spillway	Ogee gated
10.	Details of gates	
	a) Type	Radial
	b) Size in meter	11.58 X 12.49
	c) Nos.	10 Nos
11.	Cost	
	a) Irrigation + Dam	1693.978 lakhs
	B) Hydro	267.287 lakhs
	c) Total	1961 lakhs

(M.W.L–Maximum Water Level, F.R.L.- Full reservoir level, M.- Meters)

Track Map to Reach the Yeldari Reservoir

Yeldari reservoir is easiest place to reach by all kinds of vehicles; it is located 15 km. from Jintur on Jintur to Sengaon road via Yeldari village (camp).Near the embankment of Yeldari reservoir, there are four closely linked villages namely Yeldari camp, Sawangi Mahalsa and Murumkheda from Jintur taluka in Parbhani Dist. and Yeldari village from Sengaon taluka of Hingoli District. Another way to reach the Yeldari reservoir is via Sengaon-Jintur road, 30 km away from Sengaon taluka of Hingoli District. Except these tar roads other road facilities are very poor and remote.

On the Yeldari reservoir, there is a guest house of Irrigation department, in which there are two special rooms and four general rooms for accommodation of the government officials, researchers, leaders, social workers and special tourists.With the permission of chief executive engineer office at Yeldari camp one can book the room in this guest house. Fish and prawn dishes are special preparations available here from the fresh catch at Yeldari reservoir. Best season to visit the reservoir is from January to May.

Materials and Methods

For the study of Yeldari reservoir on various aspects of its fishery management as mentioned below, the data was collected by the survey and visit to the DFDO office, Parbhani (recently it is being restructured and named as, The Office of the Assistant Commissioner of Fisheries), from the fisher communities living near the embankment of the reservoir and distributed in 34 villages located around the reservoir. The data on the structure and working pattern of Fish co-operative society actively working on the reservoir was collected from the office of the society namely Purna Matsyavyvasai sahakari Sanstha, Maryadit, Sawali (Bu) by personal interviews, by using the questionnaires, photography and videography, for the duration of June 2005 to May 2008. The data was analyzed by applying various statistical tools and techniques. The result of the study were compared and discussed with the fishery management of various popular lakes and reservoirs in India and abroad.

Fisheries Management

To get the maximum fish from the reservoir is a need of the time. For high yield of fish from the reservoir the application of new techniques, identification of problems in reservoir fishery management and sustainable use of natural resources, socio-economic development of the dependent fisher community are essential considerations. However, to study the Yeldari reservoir as a case study, the following points were taken into consideration.

1. Fishing tender and tender cost.
2. Fish Seed Stocking
3. Fish Harvesting–Methods, application of fishing gears and crafts, boats, purchase and maintenance of nets and boats.
4. Fish fauna
5. Fish marketing–Collection of catch, Preservation, Transportation, and marketing.
6. Fisher communities.
   (*a*) Caste and tribe
   (*b*) Population
   (*c*) Involvement of fisher women in fishing
   (*d*) Housing
   (*e*) Educational status
   (*f*) Educational facilities.
   (*g*) Wages
   (*h*) Income
7. Modern facilities of entertainment and communication. Problems of aquatic weeds, obstacles and hide outs.
8. Problem of flood and water release from flood gates of the reservoir..
9. Success of the Yeldari reservoir fishery and future prospects.

Fishing Tender and Tender Cost

The fishing tender of Yeldari reservoir was allotted to Purna Matsyavyvasai Sahakari Sanstha Maryadit, Sawali (Bu) by the District Fisheries Development Officer, Parbhani (D.F.D.O., Parbhani) by the letter No.- 4102, dated 04/10/2005 for the duration of 5 years *i.e.* from July 2005 to June 2010 at the tender cost of Rs. 1, 41,290 per year.Fish seed stocking:

Fish Seed Stocking

In the Yeldari reservoir the young stages of fishes mainly fry, semi-fingerling and fingerling commonly called as 'fish seed' were stocked by the fishing tender owners. The fish seed of Indian Major Carps namely *Catla catla, Labeo rohita, Cirrhina mrigala, Cyprinus carpio* and exotic carp species *Ctenopharyngodon idella* and *Hypopthalmichthys molitrix* were regularly stocked during the month of June to September every year.

According to Maharashtra State Government decision of Animal Husbandry, Dairy and Fisheries Department, G.R.No. Fishery Dept/1999/20/L.N./8/ADF-13 on dated 15th October 2001. The maximum number of fingerlings to be stocked in Yeldari reservoir of 6272.00 ha. average water spread area are 34.30 Lakh. The fish seed was purchased from govt. fish seed hatcheries located at Siddheshwar reservoir and Bhategaon reservoir in Hingoli district and from Masoli fish seed hatchery located at Masoli reservoir near taluka Gangakhed in Parbhani district. The fish seed was also purchased from various private fish seed hatcheries and fish seed suppliers. The details of fish seed stocked in Yeldari reservoir is given in Tables 23.1, 23.2 and 23.3.

Table 23.1: Fish Seed Stocking in Yeldari Reservoir Last 10 Years

Sl.No.	Year	Number of Fish Seed Stocked	Fishing Tender Owner
1.	1998–1999	20,00,000	Sawangi Mahalsa Matsyavyvasai Sahakari Sanstha Maryadit, Sawangi Mahalsa, Tq. Jintur Dist.Parbhani
2.	1999–2000	4,17,500	— do —
3.	2000–2001	Not Stocked	Cancellation of Lease Agreement between D.F.D.O. and Sawangi Fish Co-operative Society.
4.	2001–2002	Not Stocked	Blank Year
5.	2002–2003	Not Stocked	Blank Year
6.	1/1/2003 to 31/12/2003	16,00,000	International Meretek Pvt. Ltd., Nagpur Tq. Dist. Nagpur.
7.	1/1/2004 to 31/12/2004	Not Stocked	International Meretek Pvt. Ltd., Nagpur Tq. Dist. Nagpur.
8.	2005-2006	Not Stocked	Purna Matsyavyvasai Sahakari Sanstha Maryadit, Sawali (Bu)., Tq. Jintur Dist. Parbhani working from 4/10/05
9.	2006-2007	2,54,13, 199 fish seed 60,85,911 Prawn seed	— do —
10.	2007-2008	Not Stocked	— do —

Table 23.2: Fish Fingerling Stocking and Private Agencies Supplying Fish Seed to Stock in to the Reservoir

Sl.No.	Date of Fish Seed Stocked	Private Fish Seed Supplying Agencies	Fingerlings of Fish Species Stocked	Total No. of Fish Seed Stocked
1.	10/06/2006 and 16/06/2006	S.M. Fish Seed Suppliers, Mumbai	IMC	1,92,36,745
2.	20/06/2006 and 30/07/2006	Pathak fish Suppliers Mumbai	IMC	22,75,650
3.	13/09/2006 and 24/09/2006	Bhoiraj Fish Seed Production Centre, Gangakhed, District Parbhani	IMC	3,30,000
4.	29/09/2006 and 05/10/2006	Sameer Fish Suppliers, Mumbai	IMC	2,45,114
5.	08/10/2006 and 19/10/2006	Wanraj Enterprises fish Seed Suppliers, Mumbai	IMC	12,41,770
6.	27/10/2006	Shah Aqua India Fish Seed Suppliers, Mumbai	Silver Carp	2,83,480
	Total			**2,54,13, 199**

New Trends in Fish Seed Stocking Pattern by the Purna Fish Co-operative Society

Purna Matsyavyvasai Sahakari Sanstha Maryadit, Sawali (Bu) had taken Yeldari reservoir on lease from the State Fisheries Department (DFDO, Parbhani) for the duration of five years from July 2005 to June 2010 for fishing. The Society had made an agreement with a private capital investing company in this sector. The Purna Matsyavyvasai Sahakari Sanstha Maryadit, Sawali (Bu) had made an agreement with Intellect Agri. Product Private Limited, Andheri (East), Mumbai, Maharashtra for fish seed stocking and purchase of fish and prawn catch from the reservoir on dated 27/02/06.

Table 23.3: Prawn Seed Stocking and Private Prawn Seed Supplying Agencies

Sl.No.	*Dates of Prawn Seed Stocked*	*Private Prawn Seed Supplying Agencies*	*Prawn Species Stocked*	*No. of Prawn Seed Stocked*
1.	13/09/06 and 11/11/06	Surya Prawn Seed Supliers, Mumbai	*Macrobranchium rosembergii*	11,51,220
2.	06/09/06 and 15/10/06	S.M. Prawn Seed Supliers, Mumbai	*Macrobranchium rosembergi*	23,45,116
3.	08/10/06 and 15/12/06	Wanraj Prawn Seed Supliers, Mumbai	*Macrobranchium rosembergi*	15,64,927
4.	14/10/06 and 18/10/06	Chahare Prawn Seed Supliers, Washim, Maharashtra	*Macrobranchium rosembergi*	7,72,800
5.	14/11/06	Bhagwati Fishing Corporation Ahmadabad, Gujarat	*Macrobranchium rosembergi*	41,275
6.	09/12/06 and 16/12/06	Samarth Aquaculture Seed Accessories, Mumbai.	*Macrobranchium rosembergi*	2,12,573
	Total			60,85,911

Figure 23.1: Fish Seed Stocking in Yeldari Reservoir

According to an agreement between the Fish co-operative society and the Intellect Agri. Product Private Limited Company, the company has to stock the fish fingerlings of Indian major carp species and prawn juveniles of *M. rosenbergi* species in Yeldari reservoir and have to purchase all the fish and prawn catch regularly from the fishermen involved in fishing in Yeldari reservoir and the Intellect Agri. Product Private Ltd. Company also has to give Rs. 5 / kg for fishes and Rs. 10 / kg for prawns as commission to the society.

According to agreement Intellect Agri. Products Pvt. Ltd. Company had purchased fish fingerlings and prawn juveniles from different private fish seed and prawn seed suppliers. The Purna Matsyavyvasai Sahakari Sanstha Maryadit, Sawali (Bu) (PMSSM, Sawli Bu) had kept the details of seed stocking of fish species and prawn species and the private fish seed and prawn seed supplying agencies shown in Table 23.1,23.2 and 23.3. The total no. of fingerlings stocked were 2,54,13, 199 of which Indian Major Carp fingerlings were 2,51,29,719 and silver carp fingerlings were 2,83,480. The time duration required for fingerlings stocking was from June 2006 to October 2006. The total no. of prawn juveniles of *Macrobrachium rosenbergii* species stocked in Yeldari reservoir were 60, 85,911 number, this was the first time of such heavy stocking of prawn seed in to the reservoir in the history of Yeldari reservoir fishery. The time duration required for the stocking of prawn seed was from Sept. 2006 to Dec. 2006 and fish seed was from June 2006- December 2006.

The fish seed and prawn seed were brought to the P.M.S.S.M., Sawli (Bu) Office from different places through containers loaded in trucks during the period of June 2006 to December 2006. The fish seed was released into the Yeldari reservoir only at one station *i.e.*, near the office of the P.M.S.S.M., Sawali (Bu) near the Reservoir's earthen embankment at Yeldari camp.

For the fish seed stocking the P.M.S.S.M.Sawali (Bu) has adopted different method from the routine to release the fish seed into the reservoir. The purchased fish seed was brought in to the plastic bags filled with water and Oxygen gas partially, and the bags were protected and packed in the plastic foldable tins. All the seed containing tins were loaded in to the large containers. Instead of release of fish seed by opening the fish seed bags, the fish seed bags were emptied into 6 inch diameter PVC plastic pipes connected in to the reservoir water from the embankment of reservoir. It was probably due to avoid the labor cost and to save the time. It was a successful method to stock the huge quantity of fish seed in crores of number for the large reservoirs like Yeldari.

In the year 2006-07, the flood gates of Yeldari reservoir were opened for 3 times during the period of August 2006 to October 2006 and were not opened during November 2006 to March 2007.(Source :Sub Divisional Officer, Maintenance,Sub Division,Yeldari camp Tq. Jintur Dist.Parbhani.)

During the period of August 2006 when water was released for first time, it was considered that there may be loss of stocked fish seed from the reservoir. The fish and prawn seed was stocked in large quantity in to the reservoir to avoid the decrease in population of stocked fish seed and prawn seed loss through the released water from Yeldari reservoir in to Purna river basin, which later on enter in to the Siddeshwar reservoir.

It was found that, Intellect Agri Products Pvt. Ltd, Mumbai was not working properly according to the rules and conditions of the agreement with the Purna Fish Co-operative society.Therefore new agreement was finalized between Purna fish society and Aqua Fisheries and Agro Products Pvt. Ltd. Nariman Point, Mumbai on dated 15/02/2007 for the further period, the agreement was finalized by tender system, this agreement was with all legal deed process, It was an agreement only for the purchase of the catch from the reservoir and Co-operation in the fishery management process and the company has to pay the amount of stocked fish and prawn seed to the earlier company *i.e.* Intellect Agri Products Pvt. Ltd Company,Mumbai.Therefore earlier agreement with the Intellect Agri Products Pvt. Ltd., Mumbai was cancelled by the Purna Fish Co-operative Society.

Table 23.4: The Fish Catch at Yeldari Reservoir from June 17, 2006 to May 31, 2008

Sl.No.	Months and Year	Yeldari Fish Collection Centre		Bamni Fish Collection Centre		Khadki Fish Collection Centre		Average Fish Catch/ Day of Month in kg	Total Fish Catch of Month in kg
		Fish Catch/Day of the Month in kg		Fish Catch/Day of the Month in kg		Fish Catch/Day of the Month in kg			
		Minimum	Maximum	Minimum	Maximum	Minimum	Maximum		
1.	Jun-06	—	—	—	—	—	—		
2.	Jul-06	30	155	—	—	—	—	92.5	2867.5
3.	Aug-06	28	944	—	—	—	—	486	15066
4.	Sep-06	179	1612	—	—	—	—	895.5	26895
5.	Oct-06	21	446	—	—	—	—	233.5	7238.5
6.	Nov-06	120	483	—	—	—	—	302.5	9045
7.	Dec-06	374	674	—	—	—	—	524	16244
8.	Jan-07	338	713	—	—	—	—	525.5	16290.5
9.	Feb-07	233	612	158	342	—	—	672.5	18830
10.	Mar-07	235	674	141	441	91	332	954	29667
11.	Apr-07	45	467	67	342	95	280	648	19440
12.	May-07	60	281	20	343	104	348	578	17918
13.	Jun-07	117	598	51	254	242	369	815.5	24465
14.	Jul-07	233	928	166	732	135	755	1474.5	45709.5
15.	Aug-07	233	1013	73	606	109	788	1411	43741
16.	Sep-07	334	785	130	495	313	902	1479.5	44385
17.	Oct-07	386	1038	182	602	362	776	1673	51863
18.	Nov-07	381	1176	94	326	337	904	1609	48270
19.	Dec-07	439	1014	38	196	316	814	1408.5	43663.5
20.	Jan-08	329	1055	27	134	231	560	1168	36208
21.	Feb-08	714	1434	36	234	316	742	1788	50295
22.	Mar-08	208	1081	34	253	193	1107	1438	44578
23.	Apr-08	304	930	40	212	250	870	1303	39090
24.	May-08	210	880	70	250	210	850	1235	38285
	Total	**5551**	**18993**	**1327**	**5762**	**3304**	**10397**	**22667**	**690054.5**

Fish and Prawn Harvesting

On Yeldari reservoir there was 'no closed' day for fishing. Though P.M.S.S.M Sawli (Bu) had taken the fishing tender of Yeldari reservoir for the duration July 2005 to June 2010 On dated 4th Oct

2005. The reservoir was free for fishing from July, 2005 to June, 2006 to the Fishermen of villages located around the Yeldari reservoir; in this duration the fishermen engaged in fishing were about 250-300 in number from 34 different villages.

Table 23.5: The Prawn Catch at Yeldari Reservoir from June 17, 2006 to May 31, 2008

Sl.No.	Months and Year	Yeldari Prawn Collection Centre		Bamni Prawn Collection Centre		Khadki Prawn Collection Centre		Average Prawn Catch/ Day of Month in kg	Total Prawn Catch of Month in kg
		Prawn Catch/Day of the Month in kg		Prawn Catch/Day of the Month in kg		Prawn Catch/Day of the Month in kg			
		Minimum	Maximum	Minimum	Maximum	Minimum	Maximum		
1.	Jun-06	—	—	—	—	—	—	—	—
2.	Jul-06	1.75	5.25	—	—	—	—	3.5	52.5
3.	Aug-06	2	11	—	—	—	—	6.5	100.75
4.	Sep-06	0.25	3	—	—	—	—	1.625	24.375
5.	Oct-06	0.25	2	—	—	—	—	1.125	17.4375
6.	Nov-06	0.25	4	—	—	—	—	2.125	31.875
7.	Dec-06	1.25	6	—	—	—	—	3.625	56.1875
8.	Jan-07	0.25	4	—	—	—	—	2.125	32.9375
9.	Feb-07	0.25	1.25	0.25	1			1.375	19.25
10.	Mar-07	0.25	5	0.25	1	0.25	0.75	3.75	58.125
11.	Apr-07	0.25	21	0.75	10	0.25	30.5	31.375	470.625
12.	May-07	3	26	4	233	19	123	204	3162
13.	Jun-07	69	386	15	73	49.75	101	346.875	5203.125
14.	Jul-07	87	199	34	104	44	139	303.5	4704.25
15.	Aug-07	52	187	21	62	36.5	113	235.75	3654.125
16.	Sep-07	57	104	5	48	19	64	188.5	2827.5
17.	Oct-07	27	88	8	43	11	25	101	1565.5
18.	Nov-07	39	88	7	31	9	25	99.5	1492.5
19.	Dec-07	20	50	1	6	5	15	48.5	751.75
20.	Jan-08	14	44	0.25	5	4	16	41.625	645.187
21.	Feb-08	16	61	0.25	7	4	34	61.125	886.312
22.	Mar-08	16	62	2	29	4	33	73	1131.5
23.	Apr-08	16	60	2	25	4	30	68.5	1027.5
24.	May-08	16	50	0.5	10	4.5	30	55.5	860.25
	Total	**438.75**	**1467.5**	**101.25**	**688**	**214.25**	**779.25**	**1844.5**	**28775.5615**

Indian major carp species, the exotic carp species and *Macrobrachium rosenbergii* prawn species stocked in the Yeldari reservoir were harvested by using various kinds of nets called by various local traditional names. Gill

Fish Marketing

Marketing of Fish Catch Before June 2006

During June 2005 to May 2006 the fish catch from the Yeldari reservoir was distributed and marketed by different methods. Some of the fishermen purchase the fish catch for Rs. 30 to 40 per kg. for fishes more than 01 kg. weight and for Rs. 20 to 30 per kg. for fishes less than 1 kg. weight and they sale the fishes to middle man on commission at 5 to 10 Rs/kg. for the large sized fishes and Small sized fishes. Most of the middlemen directly purchase the fishes from fishermen on Yeldari reservoir and sale the fishes in the surrounding market places at higher rate.

Table 23.6: Marketing of Catch from Yeldari Reservoir and Wages (Year 2006-2008)

Sl.No.	*Fish/Prawn*	*Wages to Fishermen*	*Purchase Rate by Aqua Fisheries and Agro Products Pvt Ltd. from the Society*	*Purchase Rate by Middle Man from Aqua Fisheries and Agro Products Pvt Ltd*	*Purchase Rate by Retailers from Middle Man*
1.	Local fishes, fishes below 1 kg	19 Rs. / kg	24 Rs/kg	29 Rs/kg	31-34 Rs/kg
2.	Fish above 1 kg	24 Rs. / kg	29 Rs/kg	34 Rs/kg	36-48 Rs/kg
3.	Prawns (Feb. to June 07)	150 Rs./kg	NA	230 Rs/kg	250-290 Rs/kg
4.	Prawns (July 2007 onwards)	70 Rs./kg	NA	230 Rs/kg	250-290 Rs/kg

Some of the fishermen and fisherwomen working on the reservoir collect their own fish catch and catch from other fishermen and sale it in the surrounding villages and market places.

The freshly collected fishes of Yeldari reservoir were sold in Yeldari-camp, Jintur, Selu, Parbhani, Risod, Sengaon, and Washim market. The freshly collected fishes were transported from Yeldari reservoir to different market places on bicycle, motor cycle or in four wheelers like tempo.

The fish preservation technique and facilities were not found during this study period (before June 2006). Recently a cold storage plant is being constructed near the reservoir at Yeldari campin December 2007. Visit to this cold storage plant for ots detail study was not permitted by the concerned authorities. The ownership of the cold storage plant was towards Aqua Fish products Pvt. Ltd., Mumbai in collaboration with Purna Fish Co-operative Society working at the Yeldari reservoir.

Marketing of Fish Catch After June 2006

From June 2006, due to proper functioning of Purna Fish Co-operative Society, the fishery management of Yeldari reservoir was improved.

From June 2006, The society had banned the use of 'Zorli', or 'Wadap', 'Pandya' and 'Purai' type drag-bag nets and Phekjal type cast net because, by the use of these nets, the pre-mature young, small sized fishes may be harvested and it will affect negatively on the reservoir fish production. Due to the use of nets of very small mesh size of less than 1 cm., Purna society had banned the fishing during Jun, 2006 to July, 2006 in Yeldari reservoir by considering the breeding period of fishes, but in some villages located around the reservoir like Bamni, Khadki, Kini, Borkhedi the fishing was observed.

In July, 2006 Purna society had established the first fish collection centre at Yeldari-camp. All catch obtained by the fishermen was collected at Yeldari camp fish collection centre up to December 2006. In this duration, with the help of 200 fishermen the fish fingerlings and prawn juveniles were stocked by the Purna society in Yeldari reservoir near Yeldari-camp fish collection centre, the fish seed contains Indian major carps,other local fish species and *M. rosenbergii* prawn. The prawn seed stocked in some irrigation reservoirs of Sengaon taluka in Hingoli district which reached in to the Yeldari reservoir through overflow.

It was became inconvenient to bring the fish catch at Yeldari-camp fish collection centre by the fishermen of different villages except from few villages like Yeldari-camp, Sawangi Mahalsa, Murumkheda, Kini, Yeldari, Limbala Tanda, Ambarwadi and Kawtha. Mean while fish catch of Kawtha village was weighed and brought to Yeldari-camp fish collection centre by using diesel engine boat or petrol engine speed boat by the Purna Fish Co-operative Society employees.

In January, 2007 the second centre for fish and prawn collection was started at village Bamni which was 12 km. away from Yeldari camp for the Fishermen of Ambarwadi, Badnapur, Belkheda, Umrad, Saikheda, and Wazzar villages.

In February, 2007 the third centre for fish and prawn collection was started at Khadki,in Hingoli district, about 22 km away by road distance from Yeldari-camp, for the fishermen of villages Bhandari, Volgira, Khairi Ghumat, Pathonda, Borkhadi tanda, Dhotra, Sonsawangi, Khadki, Nansi, Bamni (ku), Dongaon, Salegaon, Ooty (Purna), Dhanora and Wazar of taluka Sengaon District Hingoli.

Fish and Prawn Collection

From February, 2007 onwards daily catch obtained by the fishermen was collected at 3 main fish collection centers established by the society. For the fish and prawn collection from Bamni and Khadki fish collection centers two jeeps with 5 to 6 employees from Intellect Agri Product Company in each jeep were involved. Regularly both jeeps along with employees go to Bamni fish collection centre and Khadki fish collection centre. These employees collect the fish and prawn catch from the fishermen and take the necessary records in their register of fish and prawn catch. They weigh the fish and prawn catch and pay the payment to the fishermen on the spot regularly for the catch. Later on these employees sort out the fish and prawn catch according to species and size of fishes and give it to middle man appointed by Intellect Agri Product Company. During August 2007 to November 2007 the prawn catch was not given to middle man and prawn catch was collected from Bamni and Khadki fish and prawn collection centre to Yeldari-camp fish and prawn collection centre. During this period the prawn was marketed to Mumbai, Kolkata and Amritsar fish markets. It was also reported that, the prawns of Yeldari reservoir were exported to China.

Yeldari-Camp Fish and Prawn Collection Centre

Yeldari-camp fish and prawn collection centre was present near the office of Purna Fish Co-operative Society at Yeldari-camp. From July 2006 onwards. The daily fish and prawn catch from fishermen of villages Yeldari and Limbala Tanda of Tq. Sengaon Dist. Hingoli and Yeldari camp, Sawangi Mahalsa, Murumkheda, Kini and Kawtha villages of Tq. Jintur Dist. Parbhani were purchased and payment for the catch was given on the spot to the fishermen at this centre. The fishes and prawns were sorted according to species and size and weighed and given to middle-man regularly according to the agreement between Intellect Agri Product Company and middle-men.

The middle man sale the fishes and prawns on the spot to 5 to 6 retailers for Rs. 35 to 40 per kg for local fishes and IMC fish species less than 1 kg, where as IMC fish species more than 1 kg for Rs.36 to

50 per kg. The remaining fish and prawn catch was sold to the traders at Parbhani. From Parbhani the fish and prawn catch was sold in Parbhani and surrounding markets and remaining fishes and prawns were transported to Kolkata, Mumbai and Amritsar by Railway transportation.

Bamni Fish and Prawn Collection Centre

At Bamni fish collection centre located near village Bamni, daily fish and prawn catch from the fishermen of Ambarwadi, Chaudharni, Badnapur, Bamni, kolpa, Sawangi (Bhambre), Belkheda, Umrad, Saikheda, and Wazar villages of were collected.The fishes and prawn catch were weighed and fish catch details were noted in the register maintained by the Society and payment for fishes and prawns were given on the spot by the Purna society regularly to the fishermen.

All the collected fishes and prawns were sorted and weighed by Purna Society and given to middle-man regularly according to the agreement between Intellect Agri Product Company and middle-men at the rates given in Table 3A..

The middle men sale the fishes and prawn to 5– 6 retailers at rupees 35 to 40 per kg for local fishes and IMC fish species less than 1 kg, where as IMC fish species more than 1 kg at rupees 36 to 48 per kg. The remaining fish and prawn catch was sold to the trader at Parbhani. From Parbhani the fish and prawn catch was sold in Parbhani and surrounding market and remaining fishes and prawn were transported to Calcutta, Bombay, Amritsar, via Railway transportation.

Khadki Fish and Prawn Collection Centre

At Khadki fish collection centre daily fish and prawn catch from fishermen of villages Bhandari, Volgira, Khairi Ghumat, Pathonda, Borkhadi tanda, Dhotra, Sonsawangi, Khadki, Nansi, Bamni (khu.), Dongaon, Salegaon, Ooty (Purna), Dhanora and Wazar of Sengaon taluka District Hingoli were collected. The fish catch from Bhandari, Volgira, Khairi Ghumat, Pathonda, and Borkhadi villages was collected at Bhandari and later on carried to Khadki fish and prawn collection centre. From each village, 1 to 4 fishermen regularly collect their own fish and prawns catch and catch from fishermen was already weighed and bring it on motorcycle to Khadki fish collection centre between 9 to 11 am. The fish and prawn catch of fishermen were weighed and fish catch details were noted in the register of the Purna Society and payment for fishes and prawns were given on the spot by the Purna society regularly. All the collected fishes and prawns were sorted and weighed by Purna Society and give it to middle-man regularly according to the agreement between Intellect Agri Product Company and middle-men.

The middle men collect all the fishes and prawn and preserve it in the containers with ice and they load the containers in mini truck or Jeep and carry to the Risod and Washim fish market. It is known from relevant source that, a very limited fishes were sold at Risod fish market @ Rs. 40 to 50 per kg. for local fish and @ Rs.50 to 60 per kg. for IMC fish species through agent or fish sellers. Most of the fishes and almost all prawns were sold to the fish traders at Washim @ Rs.25/kg. for local fish and @ 40 Rs./kg. for IMC species and @ Rs. 250/kg. for prawn. From Washim, the fishes were transported in local market, and the prawns were marketed at Akola, Nagpur and Mumbai markets.

Fisher Communities

Fisher Population

Around the Yeldari reservoir there are34 villages located closer to reservoir water. In these 34 villages the fisher communities of different caste and tribes are present. During the present study period the fisher population was recorded and it was 964. From almost all the 34 villages there are 16 villages included in Jintur taluka of Parbhani District where as 28 villages included in Sengaon taluka of Hingoli District located on other side of reservoir. The fisher population recorded from 16 villagesof Taluka Jintur Dist. Parbhani was 526 and from 18 villages included in Tq. Sengaon Dist. Hingoli was 438. The population status of fisher communities distributed around Yeldari Reservoir is given in the Table 23.7 and 23.8. Out of 964 fishermen, fulltime active fishermen number was 524. The fulltime active fishermen population recorded from 16 villages included in Tq. Jintur Dist.Parbhani was 343 and from 18 villages included in Tq. Sengaon Dist. Hingoli was 181. Population Status of Fulltime active Fishermen working ar Yeldari Reservoir is given in the Tables 23.7 to 23.9.

Table 23.7: Fishermen Population Status of Yeldari Reservoir (Year 2006–2008)

Sl.No.	*Village Name*	*Taluka of District Parbhani and Hingoli*	*Fishermen Population*	*Fisher Community Tribes and Caste and their Population Number*
1.	Yeldari camp	Jintur	85	Bhoi-51, Boudh-30, Telgu-1, Muslim-03
2.	Sawangi Mahalsa	Jintur	46	Banjara-03, Bhoi-09, Boudh-20, Holar-02, Muslim12
3.	Murumkheda	Jintur	08	Bhoi-02, Boudh-06,
4.	Kini	Jintur	69	Bhoi-12, Boudh-15, Chambar-36, Hatkar-01, Vanjari-05
5.	Ambarwadi	Jintur	56	Andh-2, Boudh-11, Chambar-11, Matang-23, Vanjari02, Waddar-07
6.	Kawtha	Jintur	76	Andh-48, Bhoi-03, Boudh-12, Hatkar-08, Matang-05,
7.	Badnapur	Jintur	07	Bhoi-07
8.	Chaudharni	Jintur	21	Boudh-17, Hatkar-02, Muslim-02
9.	Bamni	Jintur	32	Bhoi-06, Baudh-03, Chambar02, Dhanger-02, Hatkar-01, Koli-01, Matang10, Muslim 7
10.	Kolpa	Jintur	37	Bhoi-06, Boudh-22, Hatkar-07, Koli-02
11.	Kumbhephal	Jintur	09	Boudh-09
12.	Sawangi Bhambre	Jintur	43	Bhoi-09, Boudh-11, Koli-21, Muslim-02
13.	Umrad	Jintur	23	Boudh-23
14.	Bekheda	Jintur	03	Boudh-03
15.	Saikheda	Jintur	03	Boudh-03
16.	Wazar(Kh)	Jintur	08	Boudh-02, Rajput-05, Muslim-01
17.	Yeldari	Sengaon	10	Hatkar-10
18.	Limbala-Tanda	Sengaon	23	Banjara-23
19.	Bhandari	Sengaon	32	Andh-01, Banjara-05, Bhoi-03, Boudh-05, Hatkar-01, Muslim-17
20.	Khairi-Ghumat	Sengaon	13	Banjara-05, Hatkar-06, Muslim-02.
21.	Holgira	Sengaon	07	Banjara-05, Hatkar-02
22.	Borkhadi	Sengaon	39	Banjara-26, Boudh-12, Hatkar-01
23.	Dhotra	Sengaon	49	Boudh-41, Hatkar-07, Maratha-01
24.	Pathonda	Sengaon	26	Andh-11, Banjara-04, Boudh-10.Hatkar-01
25.	Son Sawangi	Sengaon	25	Banjara-22, Hatkar-03

Contd...

Table 23.7–Contd...

Sl.No.	Village Name	Taluka of District Parbhani and Hingoli	Fishermen Population	Fisher Community Tribes and Caste and their Population Number
26.	Khadki	Sengaon	43	Banjara-21, Boudh-21, Hatkar-01
27.	Bamni(Kh)	Sengaon	38	Banjara-31, Hatkar-03, Muslim-04
28.	Nansi	Sengaon	27	Banjara-02, Boudh-25
29.	Dongergaon	Sengaon	43	Banjara-21, Bhoi-16, Boudh-06
30.	Ooty(Purna)	Sengaon	20	Bhoi-08, Boudh-12,
31.	Salegaon	Sengaon	13	Boudh-11, Muslim-01, Vanjara-01
32.	Dhanora	Sengaon	20	Bhoi-08, Boudh-06, Hatkar-04, Maratha-02
33.	Pimpri	Sengaon	06	Boudh -06
34.	Barda	Sengaon	04	Boudh-04
Total Fishermen Population = 964				

Table 23.8: Full-time Fishermen Population Status of Yeldari Reservoir (Year 2006–2008)

Sl.No.	Village Name	Taluka	Full Time Fishermen Population	Fisher Community Tribes and Caste and their Population Number
1.	Yeldari camp	Jintur	76	Bhoi-45, Boudh26, Muslim-03, Telgu-1
2.	Sawangi Mahalsa	Jintur	38	Banjara-02, Bhoi-07, Boudh-18, Holar-01, Muslim-10,
3.	Murumkheda	Jintur	04	Bhoi-01, Boudh-03.
4.	Kini	Jintur	45	Bhoi-08, Boudh-11, Chambar-24, Hatkar-01, Vanjari-01
5.	Ambarwadi	Jintur	38	Boudh-08, Chambar-10, Matang-13, Vanjari-02, Waddar-05
6.	Kawtha	Jintur	63	Andh-41, Bhoi-02, Boudh-09, Hatkar-06, Matang-05,
7.	Badnapur	Jintur	04	Bhoi-04
8.	Chaudharni	Jintur	12	Boudh-09, Hatkar-01, Muslim-02
9.	Bamni	Jintur	12	Bhoi-02, Boudh-02, Dhangar-01, Matang-05, Muslim-02,
10.	Kolpa	Jintur	17	Bhoi-04, Boudh-09, Hatkar-03, Koli-01
11.	Kumbhephal	Jintur	04	Boudh-04
12.	Sawangi Bhambre	Jintur	16	Bhoi-04, Boudh-02, Koli-09, Muslim-01
13.	Umrad	Jintur	11	Boudh-11
14.	Belkheda	Jintur	02	Boudh-02
15.	Saikheda	Jintur	01	Boudh-01
16.	Wazar(Kh)	Jintur	10	Hatkar-10
17.	Yeldari	Sengaon	06	Hatkar-06
18.	Limbala-Tanda	Sengaon	09	Banjara-09
19.	Bhandari	Sengaon	20	Andh-01, Banjara-01, Bhoi-03, Boudh-04, Hatkar-01, Muslim-10
20.	Khairi-Ghumat	Sengaon	07	Banjara-02, Hatkar-04, Muslim-02.

Contd...

Table 23.8–Contd...

Sl.No.	Village Name	Taluka	Full Time Fishermen Population	Fisher Community Tribes and Caste and their Population Number
21.	Holgira	Sengaon	04	Banjara-03, Hatkar-01
22.	Borkhadi	Sengaon	16	Banjara-08, Boudh-08
23.	Dhotra	Sengaon	26	Boudh-24, Hatkar-01, Maratha-01
24.	Pathonda	Sengaon	10	Andh-05, Banjara-02, Boudh-02, Hatkar-1
25.	Son Sawangi	Sengaon	11	Banjara-09, Hatkar-02
26.	Khadki	Sengaon	15	Banjara-10, Boudh-05
27.	Bamni(Kh)	Sengaon	19	Banjara-15, Hatkar-02, Muslim-02
28.	Nansi	Sengaon	07	Banjar-01, Boudh-06
29.	Dongergaon	Sengaon	09	Banjara-06, Bhoi-02, Boudh-01
30.	Ooty(Purna)	Sengaon	07	Bhoi-03, Boudh-04
31.	Salegaon	Sengaon	01	Boudh-01
32.	Dhanora	Sengaon	13	Bhoi-07, Boudh-03, Maratha-02, Hatkar-01
33.	Pimpri	Sengaon	–	–
34.	Barda	Sengaon	–	–

Total Full Time Fishermen Population = 533

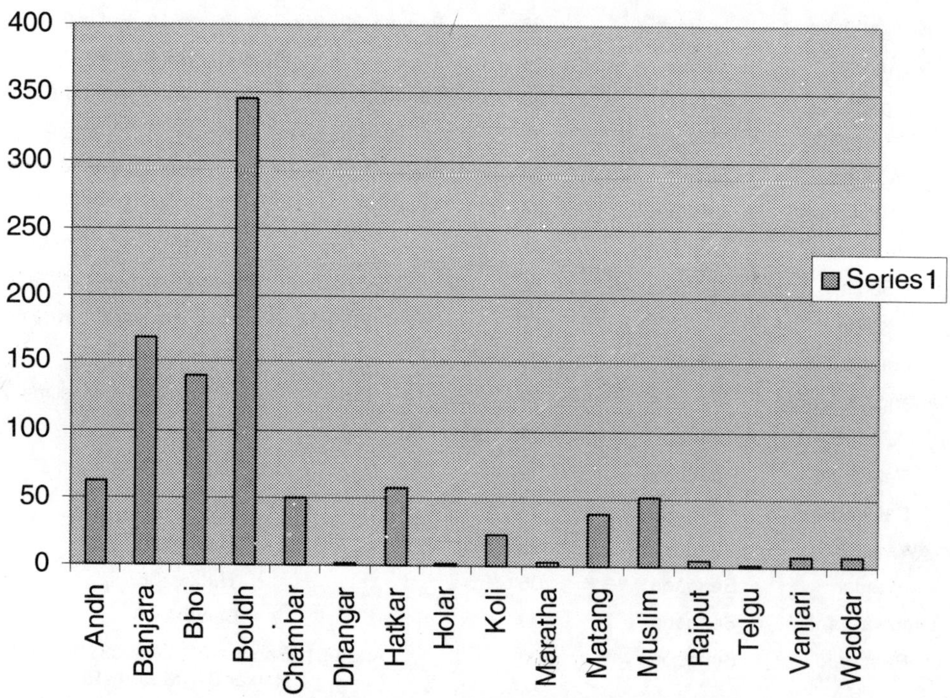

Figure 23.2: Caste and Tribe Fisher Population Number Involved in the Yeldari Reservoir Fishery (Year 2006-2008)

Caste and Tribe

The fisher population present in 34 villages belongs to different caste and tribes. The fishermen caste and tribes were Andh(Adiwasi), Banjara, Bhoi, Chambhar, Dhangar, Hatkar, Holar, Koli, Maratha, Matang, Muslim, Rajput, Telgu, Vanjari and Waddar. Details of active fishermen caste and tribe population is given in the Table 23.9.

Table 23.9: Caste and Tribe Population of Fisher Communities in the Villages around Yeldari Reservoir

Sl.No.	*Caste and Tribe of Fisher Population*	*Fisher Population of 16 Villages of Tq. Jintur Dist. Parbhani*	*Fisher Population of 18 Villages of Tq. Sengaon Dist. Hingoli*	*Total Fisher Population*	*Percentage of Population*
1.	Andh	50	12	62	6.43
2.	Banjara	03	165	168	17.42
3.	Bhoi	105	35	140	14.52
4.	Boudh	187	159	346	35.89
5.	Chambar	49	—	49	5.08
6.	Dhangar	02	—	02	0.20
7.	Hatkar	19	39	58	6.01
8.	Holar	02	—	02	0.20
9.	Koli	24	—	24	2.48
10.	Maratha	—	03	03	0.311
11.	Matang	38	—	38	3.94
12.	Muslim	27	24	51	5.29
13.	Rajput	05	—	05	0.51
14.	Telgu	01	—	01	0.10
15.	Vanjari	07	01	08	0.82
16.	Waddar	07	—	07	0.72
	Total	**526**	**438**	**964**	

Involvement of Fisherwomen in Fishing and Other Activities

It was found that the fisher women of various fisher caste and tribe are involved in the fishing activities along with their husband. At Yeldari reservoir, 8 to 10 fisher women were involved in actual fishing activities; they operate the thermocoel rafts, gill-net and also assist the fishermen in the operation of large nets like Zorli' and 'Pandya'. All fisherwomen were active swimmers. The clothing of the fisherwomen were saries, locally it was called as 'Lugda'.

Majority of fisherwomen were involved in household activities, but at some places like Yeldari-camp, Kini, Kawtha, and Ambarwadi, Bamni (Bu) of Tq. Jintur Dist. Parbhani and Bhandari, Kawtha, Bamni of Tq. Sengaon Dist. Hingoli the fisher women belongs to caste Boudh, Bhoi, Chambar, and Banjara were involved in fish marketing of their own catch and the marketing of purchased catch.

There was no remarkable involvement of fisherwomen in the Yeldari reservoir fishery but they play an important role to assist their husband in the fishery activities, but whatever fisherwomen population was actively working and involved in actual fishing activities, their activities were really brave in such a vast spread and deep reservoir, it was beyond the imagination of common man.

Housing

Around the Yeldari reservoir there are few villages around the peripheral region of the reservoir like Kini, Kawtha, and Sawangi–Mahalsa having 100 per cent population of fisher communities; in these villages, there are separate colonies of Boudh, Muslim, Chambar, and Bhoi. The type of houses of fisher communities distributed in various villages around Yeldari reservoir are of simple type, constructed with the stone, bricks, clay and concrete. Maximum houses are of tin protection, some are having concrete slabs and others are with hut like structure. All housing system is in small plot of 1500 to 2000 sq. ft. area.

The fisher communities built their temporary hut like structure of polythene and bamboo at the site of fishing away from their village on the coast of reservoir. These huts are temporary, easy to construct and dismental. The temporary huts constructed during peak fishing season. In these temporary huts sometimes entire family of fishers take the rest or only fishermen use this hut as a shelter while the rest of family members remain in their villages.

Wages and Income

The main source of income for the fisher communities working at the Yeldari reservoir can be explained as-

Situation Before June 2006

There was no fixed and assured income for the fisher community because, the Yeldari reservoir fishery management was very poor before year 2006, in this period the fishing contract was toward Meretek Company, Nagpur. Some time the reservoir remain open hence there was no ban on the use of small mesh size nets and mosquito nets for fishing, similarity the fish seed stocking by fishing tender owner companies, co-operative societies and Government agencies was inadequate as compared to the vast area of reservoir.

The fisher community's income was indefinite and poor, it was up to Rs. 50 to 60 per day per fisherman.

Situation After June 2006

After year 2006, the Purna society had taken the charge of fishing at Yeldari reservoir by tender system and flourished as a modern fishery co-operative society with new trends of reservoir fishery management. The society had stocked nearly 2.5 crores fish seed of Indian major carps and 60 lakhs juveniles of prawn species *Macrobrachium rosenbergii.* The society has also established a separate patrolling and vigilances squad on the land and in the water to control fish poaching, theft and control on the use of mosquito nets. This unit was having jeeps, petrol speed boat, binoculars, mobile sets. The result of this management and control on the young fish catch, theft of fish and fish marketing regulation finally resulted in the development of large sized fishes in the reservoir. This results in increase in the rate of fish and prawn catch to increase the wages to the fisher communities up 100-150 per days per person as assured income source in the peak fishing season. Due to this increase in wages the fishermen population involvement found increased in fishing activities after year 2006.

Educational Facilities and Educational Status

Marathi is the main language of communication and learning in the villages present around the Yeldari reservoir.Information about educational facilities and educational status of active fishermen, fisherwomen and their children were collected by survey and personal interviews of fisher communities belongs to village Yeldari camp, Sawngi mahalsa, Kini, Kawtha, Ambarwadi, Bamni and Khadki. The educational facilities present in the villages present around Yeldari reservoir are of moderate type. In most of the villages Marathi medium primary schools are present only at Yeldari-camp, Ambarwadi, and at village Bamni high school is present, where as Junior College and Senior Colleges are not present in any village present around Yeldari reservoir. For junior college and senior college education is available at town places like Jintur, Sengaon, and Parbhani etc. As the fishermen are engaged in fishing through out the day, there was not much attention and awareness about education to their children was found in all the fisher communities. The children of fisher communities belong to different caste and tribe go to the primary school. The drop out percentage of children from schools is comparatively more in fisher community than other communities present in the villages present around the Yeldari reservoir. To study the educational status 408 active fishermen from above mentioned villages were interviewed, Out of 408 active fishermen 164 *i.e.* 40.19 per cent active fishermen were illiterate and 245 *i.e.* 49.80 per cent active fishermen were literate. Out of 245 literate active fishermen 184 active fishermen *i.e.*73.77 per cent active fishermen had taken the primary education only and most of them had taken the education up to 2nd standard only, 48 active fishermen were educated up to high school, 11 *i.e.*4.50 per cent fishermen were educated up to 11th and 12th class and only 3 *i.e.*, 1.22 per cent active fishermen were graduate (Table 23.10). The educational status in fisherwomen is very worst. out of 293 fisherwomen from above mentioned village, it was found that 237 *i.e.* 80.88 per cent fisher women were illiterate and 56 *i.e.* 19.12 per cent fisherwomen were literate and educated up to primary school (Table 23.11). Out of 177 fisher community children 41 *i.e.*, 23.13 per cent children were illiterate and 136 *i.e.*, 76.83 per cent children were literate. Out of 136 literate children 113 *i.e.*83.08 per cent child were learning in primary school, 13 *i.e.*9.55 per cent children were learning in high school and 10 *i.e.* 7.35 per cent children were learning at 11th and 12th level. Out of 41 illiterate children most are girls (Table 23.12).

Table 23.10: Educational Status of Active Fishermen from some Villages Around Yeldari Reservoir

Sl.No.	Name of Village	Fisher Population	Illiterate	Literate			
				1st to 7th	8th to 10th	11th to 12th	Graduation
1.	Yeldari-camp	85	48	15	13	06	02
2.	Sawangi Mahalsa	46	24	13	8	—	—
3.	Kini	69	18	39	10	02	—
4.	Kawtha	76	24	45	07	—	—
5.	Ambarwadi	56	19	29	05	02	01
6.	Bamni	33	16	14	02	01	—
7.	Khadki	43	15	25	03	—	—
		408	164	180	48	11	3

Modern Facilities

The fisher community were having colour television set in their home, 10 to 20 per cent fisher were having personal mobile phone sets. It was observed that the use of motorcycle by the fishermen was found increased after year 2006. The motorcycles were used in fish transportation from site of catch to various fish and prawn collection centres of Purna society. At Yeldari-camp, Kini, Kawtha and Bhandari villages the use of motorcycle by fishermen was found increased.

Table 23.11: Educational Status of Fisherwomen from some Villages Around Yeldari Reservoir

Sl.No.	Name of Village	Fisher Population	Illiterate	Literate			
				1st to 7th	8th to 10th	11th to 12th	Graduation
1.	Yeldari-camp	56	45	11	—	—	—
2.	Sawangi Mahalsa	28	19	09	—	—	—
3.	Kini	46	40	06	—	—	—
4.	Kawtha	59	45	14	—	—	—
5.	Ambarwadi	47	41	6	—	—	—
6.	Bamni	21	16	05	—	—	—
7.	Khadki	36	31	05	—	—	—
		293	237	56			

Table 23.12: Educational Status of Children of Fisher Communities from some Villages Around Yeldari Reservoir

Sl.No.	Name of Village	Fisher Population	Illiterate	Literate			
				1st to 7th	8th to 10th	11th to 12th	Graduation
1.	Yeldari-camp	30	08	18	02	02	—
2.	Sawangi Mahalsa	22	06	12	02	02	—
3.	Kini	35	10	16	04	05	—
4.	Kawtha	27	05	18	03	01	—
5.	Ambarwadi	27	04	21	02	—	—
6.	Bamni	12	—	12	—	—	—
7.	Khadki	24	08	16	—	—	—
		177	41	113	13	10	—

Problems of Aquatic Weeds, Obstacles and Hide Out

It was observed that, there were no obstacles of weeds and tree boulders acting as obstruction for gill net fishing but the fishermen are well acquainted with the problem and information of these problematic areas of fishing was found transferred from senior fishermen to new comers. Hence this doesn't affect the fishing process in major.

The Yeldari reservoir fishery remains open through out year and no closing day of fishing hence there is continuous movement of the fishermen and their fishing activities in the reservoir, hence no any development of the weed species found in the reservoir.

Recently it was observed that the area near the concrete embankment of the Yeldari reservoir, acting as fish hide out because there is ban on fishing activity in 200 meter area from the embankment gates of the reservoir. This act as a safe place for fishes to hide out. This fishing restriction was from the Irrigation department to avoid any accident and planned mishap of damage of the concrete embankment like bomb explosion of the embankment of the reservoir.

If, with planned, secured and monitored fishing permission are given to the trained fishermen or if the irrigation department it self involved in fishing process, then there is a chance of getting huge fish catch from this thick populated area. It was found that, there are huge shoals of large fishes swimming near the embankment of the reservoir, which can be observed even from the concrete platform of the reservoir.

Flood and Water Release from Flood Gates

During heavy rainy season, it is essential to release the water beyond the storage capacity of reservoir, similarly for the irrigation to agriculture the water from Yeldari reservoir is released in the down stream Purna river basin and the water reach into Siddheshwar reservoir in the down stream area, from where the water is released through the canal to agriculture, through this system of water release from Yeldari reservoir the stocked young fish and prawn juveniles, large grown fishes and prawns lost because there is no any prevention or protection to prevent this loss. Recently purna society identified this problem and diploid active fishermen for the catch of lost fishes and prawn in the down stream purna river basin.

To compensate the problem of fish loss from the reservoir, the Purna Society decided to stock heavy number of fish seed in lakhs.

There is a chance of prevention of fish and prawn loss from the reservoir by constructing a well designed filter in the river basin in a suitable area.

Success of the Yeldari Reservoir Fishery and Future Prospects

The Yeldari reservoir fishery has been studied up to some extent by Sakhare (2001) especially for the physicochemical characters of reservoir water and discussed about some aspects of fishery management and concluded that the Yeldari reservoir has good fishery potential and suggested regular studies on various aspects for the fishery development.

Through this detailed study as a Yeldari reservoir case study, the real problems of fishery management, role of co-operative society and status of fisher communities, Adoption of modern techniques of reservoir fishery management by Purna Society etc are studied in detail as above and present status of Yeldari reservoir fishery is explained.

The striking change in the increase in fish production from the reservoir and increase in wages and income to fisher community was seen due to the efforts of Purna Matsyavyvasai Sahakari Sanstha Maryadit, Sawali (Bu).

The society had identified the fact that, it is necessary to stock the fish and prawn seed in the ratio of available area.The society had also employed a special team of workers as guards specially for patrolling around the reservoir for the prevention of theft, poaching and control on netting of fish and prawn from the reservoir.

If the fish co-operative societies or companies with some other new successful management plans and strategies are permitted in the Yeldari reservoir fishery then there is a better chance of Yeldari reservoir improvement in terms of increase in fish production, socio-economic upliftment of fisher communities, employment generation to trap the untamed huge fishery potential of Yeldari reservoir in future, which will help to improve the inland fishery production of this region.

References

Sakhare, V.B., 2007. *Reservoir Fisheries and Limnology*. Narendra Publishing House, Delhi.

Sakhare, V.B., 2003. Studies on some aspects of fisheries management of Yeldari reservoir, Maharashtra. *Ph.D. Thesis*, Swami Ramanand Teerth Marathwada University, Nanded.

Previous Volumes

— Volume 1 —

2007, xvi+194p., figs., tabls., ind., 25 cm Rs. 950

ISBN 81-7035-483-8

— Volume 2 —

2008, xvi+143p., col. plts., figs., tabls., ind., 25 cm **Rs. 750**

ISBN 81-7035-559-5

— Volume 3 —

2010, xiv+176p., col. plts., figs., tabls., ind., 25 cm **Rs. 800**

ISBN 978-81-7035-633-2

Index